Grundlehren der mathematischen Wissenschaften 221

A Series of Comprehensive Studies in Mathematics

David Mumford

Algebraic Geometry I
Complex
Projective Varieties

Springer-Verlag
Berlin Heidelberg New York 1976

David Mumford
Harvard University, Cambridge, Massachusetts 02138, USA

AMS Subject Classification (1970): 14 xx

ISBN 3–540–07603–4 Springer-Verlag Berlin Heidelberg New York
ISBN 0–387–07603–4 Springer-Verlag New York Heidelberg Berlin

Library of Congress Cataloging in Publication Data. Mumford, David. Algebraic geometry I (Grundlehren der mathematischen Wissenschaften; 221). Bibliography: p. Includes index. 1. Geometry, Algebraic. 2. Algebraic varieties. I. Title. II. Series: Die Grundlehren der mathematischen Wissenschaften in Einzeldarstellungen; 221. QA564.M843. 516′.353. 75–45192

© by Springer-Verlag Berlin Heidelberg 1976
Printed in Germany
Photosetting: Thomson Press (India) Limited, New Delhi
Printing and binding: Konrad Triltsch, Würzburg

Table of Contents

Introduction

Let me begin with a little history. In the 20th century, algebraic geometry has gone through at least 3 distinct phases. In the period 1900–1930, largely under the leadership of the 3 Italians, Castelnuovo, Enriques and Severi, the subject grew immensely. In particular, what the late 19th century had done for curves, this period did for surfaces: a deep and systematic theory of surfaces was created. Moreover, the links between the "synthetic" or purely "algebro-geometric" techniques for studying surfaces, and the topological and analytic techniques were thoroughly explored. However the very diversity of tools available and the richness of the intuitively appealing geometric picture that was built up, led this school into short-cutting the fine details of all proofs and ignoring at times the time-consuming analysis of special cases (e.g., possibly degenerate configurations in a construction). This is the traditional difficulty of geometry, from High School Euclidean geometry on up. In the period 1930–1960, under the leadership of Zariski, Weil, and (towards the end) Grothendieck, an immense program was launched to introduce systematically the tools of commutative algebra into algebraic geometry and to find a common language in which to talk, for instance, of projective varieties over characteristic p fields as well as over the complex numbers. In fact, the goal, which really goes back to Kronecker, was to create a "geometry" incorporating at least formally arithmetic as well as projective geometry. Several ways of achieving this were proposed, but after a somewhat chaotic period in which communication was difficult, it seems fair to say that Grothendieck's "schemes" have become generally accepted as providing the most satisfactory foundations. In the present period 1960 on, algebraic geometry is growing rapidly in many directions at once: to a deeper understanding of geometry in dimensions higher than 2, especially their singularities, and the theory of cycles on them; to uncovering the astonishing connections between the topology of varieties and their Diophantine properties (their rational points over finite fields and number fields); and to the theory of moduli, i.e., the parameters describing continuous families of varieties.

To acquire a good understanding of modern algebraic geometry, the insights of each of these periods have to be studied. In particular, it is necessary to know something both of classical projective geometry, of curves and surfaces in complex projective space and the "synthetic" tools for manipulating them (such as linear systems)—this amounts to what people call "geometric intuition"—and to know something of the analogies between arithmetic and geometry, of "Spec" and of "specialization mod p". Moreover, it is necessary to know both how algebraic and differential topology and complex analytic tools (such as Hodge theory) apply to complex varieties; and to know how commutative algebra can be used. It is not clear where to start! I have given introductory lectures on algebraic geometry on at least 5 separate occasions. I have taken a different tack each time, and there are several other approaches I would like to try in the future. This book grew out of notes that went through several nearly total revisions as a consequence. In the end, I found it impractical to teach classical geometry and schemes at the same time. Therefore, the present volume, which is the first of several, introduces only complex projective varieties. But, as a consequence, we can study these effectively with topological and analytic techniques without extensive preliminary work on "foundations". My goal is precisely to convey some of the classical geometric ideas and to get "off the ground": in fact, to get to the 27 lines on the cubic—surely one of the gems hidden in the rag-bag of projective geometry. The next volume will deal with schemes, including cohomology of coherent sheaves on them and applications, e.g., to π_1 of curve. The pedagogical difficulty here is that the definition itself of schemes is hard to swallow; and technically a massive amount of commutative algebra is needed to get schemes off the ground. My hope is that a previous acquaintance with complex projective varieties provides motivation and intuition for schemes.

A detailed list of prerequisites for this book follows this introduction. I hope it is almost entirely a subset of the list of "standard results" which are generally common property to all pure mathematicians. My goal has been not only to write a text for graduate students but to open the subject to specialists in other areas. Algebraic geometry is a subject that thrives on exchanging ideas and not on isolation and should be more universally understood! In particular, I have tried to write a book which you can browse in as well as read linearly. For some general guide to other literature, let me mention:

   i) for the 1900–1930 phase, Severi [1]*, Semple-Roth [1], and Zariski [1]

   ii) for the foundational phase, Grothendieck's tome [EGA] is standard for schemes but very hard to read. Another classic is Samuel [1].

   iii) among recent books, there is Serre [1], and Šafarevič [2]. An introductory book by R. Hartshorne is expected. An excellent survey of recent research is the publication AMS [1].

Tenants Harbor,
August, 1975   David Mumford

---

*All references are given in full in the bibliography.

Prerequisites

Algebraic geometry is not a "primary" mathematical subject, i.e., one which one builds directly from a small and elegant set of axioms or definitions. This makes it very hard to write an introductory book accessible to the 1st year graduate student. In general, this book is aimed at 2nd year students or anyone with at least some basic familiarity with topology, differential and analytic geometry, and commutative algebra. I want to list here all the concepts and theorems which will be assumed known at one point or another in this book (except for the results used only in the more difficult Appendix to Chapter 6).

I. *Topology:*

a) Besides standard elementary point set topology, the concept of a covering space is frequently used, e.g., in §§3A, 4B, 7D and 8D.

b) The classification of compact, orientable surfaces by the "number of handles" is used in §7B.

c) The homology groups (singular homology) come in twice: in §5C, in dealing with minimal submanifolds in $\mathbb{P}^n$, and in §7B, in calculating via Euler characteristic the number of handles of a smooth algebraic curve regarded as a topological space.

II. *Differential Geometry:*

Basic advanced calculus knowledge of differential forms and Stoke's theorem is used in 2 ways:

a) We assume DeRham's theorem that the periods of integrals give a perfect duality between real homology and closed forms mod exact forms in §5C.

b) We deduce from Stoke's theorem the basic properties of the residue at a pole of an analytic 1-form on a 1-dimensional complex manifold in §7C.

III. *Analytic Geometry:*

a) In §1B, we recall very quickly the definition of complex manifold, and we assume known the implicit function theorem for analytic functions. For our purposes, an analytic function is by definition a function given locally by a convergent power series of several complex variables.

b) In §4A, and §4B, we use repeatedly the fundamental local fact of analytic geometry: the Weierstrass Preparation theorem. Since this may be used to deduce the implicit function theorem, this includes the assumptions just above.

c) Once in §4A we use the fact that a complex-valued function f with $\mathrm{Re}(f)$ and $\mathrm{Im}\, f$ differentiable, satisfying the Cauchy-Riemann equations, is analytic.

A good reference for all this material is Gunning-Rossi [1].

IV. *Commutative Algebra:*

I wish I were able to cut down these prerequisites considerably, but I have been drawn into assuming the following:

a) from field theory, the concept of transcendence degree and its connection with derivations,

b) from general ring theory, the concepts of localizing a ring and a module, the concept of local ring, graded ring and graded module and the concept of the completion of a local ring. Integral dependence and integral closure are *not* used, except in a minor digression in §6C giving a second proof of one theorem.

c) Resultants are used in §2C and §4A to give elementary constructive proofs of several theorems. The basic facts about the resultant are summarized in §2C.

d) The decomposition theorem of ideals in noetherian rings is used in §1A and §4B in its really elementary form: if $\mathfrak{A} = \sqrt{\mathfrak{A}}$, then $\mathfrak{A} = \mathfrak{P}_1 \cap \ldots \cap \mathfrak{P}_s$, $\mathfrak{P}_i$ prime.

e) Several explicit facts about the formal power series ring $\mathbb{C}[[X_1,\ldots,X_n]]$ are used in Chapter I: that it is a UFD, and the formal implicit function theorem. This last is a very elementary special case of the formal Weierstrass Preparation theorem and could really be "left to the reader" as an exercise.

f) Finally, we have assumed in §§1A, 1C and 7B Krull's theorem: that if R is a noetherian local ring, $M \subset R$ its maximal ideal, $I \subset R$ any ideal, then

$$\bigcap_{n=1}^{\infty} (I + M^n) = I$$

This can however be easily deduced from the ideal decomposition theorem (see $Z - S$, vol. I, p. 217). If

$$\hat{R} \underset{\text{def}}{=} \varprojlim_{n} R/M^n$$

is the completion of R, then the theorem is equivalent to saying:

$$R \cap I \cdot \hat{R} = I.$$

It would have been nice to avoid using this somewhat less generally known result: but I don't know any straightforward way of proving Theorem (1.16) without it.

My standard reference for commutative algebra is $Z - S$, (both volumes), which was, in fact, written to be background for a book on algebraic geometry.

Chapter 1. Affine Varieties

§1A. Their Definition, Tangent Space, Dimension, Smooth and Singular Points

The beginning of the whole subject is the following definition:

(1.1) Definition. *A closed algebraic subset X of $\mathbb{C}^n$ is the set of zeroes of a finite set of polynomials $f_1, \ldots, f_m$, i.e., the set of all $x = (x_1, \ldots, x_n)$ such that $f_i(x) = 0, 1 \leq i \leq m$. We denote X by $V(f_1, \ldots, f_n)$.*

If $\mathfrak{A} = (f_1, \ldots, f_m)$ is the ideal in $\mathbb{C}[X_1, \ldots, X_n]$ generated by $f_1, \ldots, f_m$, then the set of zeroes of the f_i is also the set of zeroes of every $g \in \mathfrak{A}$, so we will denote X also by $V(\mathfrak{A})$. Note that there would be no point in definition (1.1) in using infinite sets of f_i's, since the ideal they generate would also be generated by a finite subset of them according to Hilbert's basis theorem; hence the set of zeroes of all the f_i's would equal the set of zeroes of this finite subset. We get immediately the following properties

a) $\mathfrak{A}_1 \subseteq \mathfrak{A}_2 \Longrightarrow V(\mathfrak{A}_1) \supseteq V(\mathfrak{A}_2)$,

b) $V(\mathfrak{A}_1) \cup V(\mathfrak{A}_2) = V(\mathfrak{A}_1 \cap \mathfrak{A}_2) = V(\mathfrak{A}_1 \cdot \mathfrak{A}_2)$

c) $V(\sum_{\alpha \in I} \mathfrak{A}_\alpha) = \bigcap_{\alpha \in I} V(\mathfrak{A}_\alpha)$,

d) If $\mathfrak{M}_x =$ the maximal ideal $(X_1 - x_1, \ldots, X_n - x_n)$, then $V(\mathfrak{M}_x) = \{x\}$.

As a consequence of (b) and (c), the subsets $V(\mathfrak{A})$ of $\mathbb{C}^n$ satisfy the axioms for the closed sets of a topology. The topology of $\mathbb{C}^n$ with these as closed sets will be called the *Zariski topology* as opposed to the usual topology.

e) If $\sqrt{\mathfrak{A}} = \{f \in \mathbb{C}[X] \mid f^m \in \mathfrak{A}, \text{ some } m \geq 1\}$ is the so-called radical of $\mathfrak{A}$, then

$$V(\sqrt{\mathfrak{A}}) = V(\mathfrak{A}).$$

Now according to a standard result in the theory of noetherian rings (cf. $Z - S$, vol. 1, p. 209) an ideal which equals its own radical is a finite irredundant intersection of prime ideals in a unique way:

(*) If $\mathfrak{A} = \sqrt{\mathfrak{A}}$, then $\mathfrak{A} = \mathfrak{P}_1 \cap \ldots \cap \mathfrak{P}_k$, where $\mathfrak{P}_i \not\supseteq \mathfrak{P}_j$ if $i \neq j$. This proves:

f) For any ideal $\mathfrak{A}$, let $\sqrt{\mathfrak{A}} = \mathfrak{P}_1 \cap \ldots \cap \mathfrak{P}_k$, then

$$V(\mathfrak{A}) = V(\mathfrak{P}_1) \cup \ldots \cup V(\mathfrak{P}_k).$$

(1.2) Definition. *A closed algebraic set $V(\mathfrak{P})$, where $\mathfrak{P}$ is a prime ideal, is called an affine variety.*

Examples: a) if $f \in \mathbb{C}[X_1, \ldots, X_n]$ is an irreducible polynomial, then the principal ideal (f) is prime, hence $V(f)$ is a variety. Such varieties are called *hypersurfaces*.

b) let $g_2,\ldots,g_n \in \mathbb{C}[X_1]$ be polynomials. Consider the set

$$X = \{(a, g_2(a), \ldots, g_n(a)) \mid a \in \mathbb{C}\}.$$

Then $X = V(X_2 - g_2(X_1), \ldots, X_n - g_n(X_1))$. The ideal $\mathfrak{A} = (X_2 - g_2, \ldots, X_n - g_n)$ is prime since it is the kernel of the homomorphism

$$\mathbb{C}[X_1,\ldots,X_n] \longrightarrow \mathbb{C}[X_1]$$

$$X_1 \longrightarrow X_1$$

$$X_i \longrightarrow g_i(X_1), \quad i \geq 2.$$

Therefore X is a variety. It is a simple type of *rational space curve*.

c) Let $l_1,\ldots,l_k$ be independent linear forms in $X_1,\ldots,X_n$. Let $a_1,\ldots,a_k \in \mathbb{C}$. Then $X = V(l_1 - a_1, \ldots, l_k - a_k)$ is a variety, called a *linear subspace* of $\mathbb{C}^n$ of dimension $n - k$.

A basic idea in the classical theory is the following:

(1.3) **Definition.** *Let $k \subset \mathbb{C}$ be a subfield, and let $\mathfrak{P}$ be a prime ideal. A k-generic point $x \in V(\mathfrak{P})$ is a point such that every polynomial $f(X_1,\ldots,X_n)$ with coefficients in k that vanishes at x is in the ideal $\mathfrak{P}$, hence vanishes on all of X.*

Example: In example (b) above if the coefficients of the g_i are in $\mathbb{Q}$, the point $(\pi, g_2(\pi), \ldots, g_n(\pi))$ is a $\mathbb{Q}$-generic point of this rational curve.

(1.4) **Proposition.** *If $\mathbb{C}$ has infinite transcendence degree over k, then every variety $V(\mathfrak{P})$ has a k-generic point.*

Proof. Let $f_1,\ldots,f_m$ be generators of $\mathfrak{P}$. We may enlarge k if we wish by adjoining the coefficients of all the f_i without destroying the hypothesis. Let

$$\mathfrak{P}_0 = \mathfrak{P} \cap k[X_1,\ldots,X_n]$$

and let

$$L = \text{quotient field of } k[X_1,\ldots,X_n]/\mathfrak{P}_0.$$

Then L is an extension field of k of finite transcendance degree. But any such field is isomorphic to a subfield of $\mathbb{C}$: i.e., $\exists$ a monomorphism ϕ

If $\overline{X}_i = $ image of X_i in L and $a_i = \phi(\overline{X}_i)$, I claim $a = (a_1,\ldots,a_n)$ is a k-generic point. In fact, $f_i \in \mathfrak{P}_0$, $1 \leq i \leq k$, hence $f_i(\overline{X}_1,\ldots,\overline{X}_n) = 0$ in L. Therefore $f_i(a_1,\ldots,a_n) = 0$ in $\mathbb{C}$ and a is indeed a point of X. But if $f \in k[X_1,\ldots,X_n]$ and $f \notin \mathfrak{P}$, then $f \notin \mathfrak{P}_0$, hence $f(\overline{X}_1,\ldots,\overline{X}_n) \neq 0$ in L. Therefore $f(a_1,\ldots,a_n) = \phi(f(\overline{X}_1,\ldots\overline{X}_n)) \neq 0$ in $\mathbb{C}$.

QED

For any subset $S \subset \mathbb{C}^n$, let $I(S)$ be the ideal of polynomials $f \in \mathbb{C}[X_1,\ldots,X_n]$

that vanish at all points of S. Then an immediate corollary of the existence of generic point is:

(1.5) **Hilbert's Nullstellensatz.** *If $\mathfrak{P}$ is a prime ideal, then $\mathfrak{P}$ is precisely the ideal of polynomials $f \in \mathbb{C}[X_1, \ldots, X_n]$ that vanish identically on $V(\mathfrak{P})$, i.e., $\mathfrak{P} = I(V(\mathfrak{P}))$. More generally, if $\mathfrak{A}$ is any ideal, then $\sqrt{\mathfrak{A}} = I(V(\mathfrak{A}))$.*

Proof. Given any $f \in \mathbb{C}[X]$, let k be a finitely generated field over $\mathbb{Q}$ containing the coefficients of f and let $a \in V(\mathfrak{P})$ be a k-generic point. If $f \notin \mathfrak{P}$, then $f(a) \neq 0$ hence f does not vanish identically on $V(\mathfrak{P})$; the 2nd assertion reduces to the 1st by means of (f) on p. 2.

(1.6) **Corollary.** *There is an order-reversing bijection between the set of ideals $\mathfrak{A}$ such that $\mathfrak{A} = \sqrt{\mathfrak{A}}$ and the closed algebraic subsets $X \subset \mathbb{C}^n$ set up by $\mathfrak{A} \longrightarrow V(\mathfrak{A})$ and $X \longrightarrow I(X) = $ (ideals of functions zero on X). In this bijection, varieties correspond to prime ideals and are precisely the closed algebraic sets which are irreducible (i.e., not the union of two smaller closed algebraic sets).*

(1.7) **Corollary.** *If $X = V(\mathfrak{P})$ is a variety, the ring $\mathbb{C}[X_1, \ldots, X_n]/\mathfrak{P}$ is canonically isomorphic to the ring of functions $X \longrightarrow \mathbb{C}$ which are restrictions of polynomials. This ring is called the affine coordinate ring of X and will be denoted R_X.*

The Nullstellensatz usually found in the literature applies to varieties over any algebraically closed groundfield k and is much harder to prove than (1.5).

Our main goal in this section is to give a first idea of the structure of affine varieties. Since the simplest type of varieties are the linear ones, we can try to approximate a general variety by a linear one:

(1.8) **Definition.** *Let $X = V(\mathfrak{P})$ be a variety and let $a = (a_1, \ldots, a_n) \in X$. The Zariski tangent space to X at a is the linear subspace of $\mathbb{C}^n$ defined by*

$$\sum_{i=1}^{n} \frac{\partial f}{\partial X_i}(a) \cdot (X_i - a_i) = 0, \qquad all\ f \in \mathfrak{P}.$$

We denote this space by $T_{X,a}$.

Note that for each $k \in \mathbb{Z}$, $\{a \in X \mid \dim T_{X,a} \geq k\}$ is a Zariski closed subset of X. In fact, if $f_1, \ldots, f_l$ are generators of $\mathfrak{P}$:

$$\dim T_{X,a} = n - rk\left(\frac{\partial f_i}{\partial X_j}(a)\right)_{\substack{1 \leq i \leq l \\ 1 \leq j \leq n}}$$

hence

$$\{a \in X \mid \dim T_{X,a} \geq k\} = V\left(\mathfrak{P} + \text{ideal of } (n-k+1) \times (n-k+1) - \text{minors of } (\partial f_i/\partial X_j)\right)$$

More succinctly, we can say that $\dim T_{X,a}$ is an upper semicontinuous function of a in the Zariski topology. (1.8) treats the tangent space *externally*, i.e., as a subspace of $\mathbb{C}^n$. But if we regard $T_{X,a}$ as an abstract vector space with origin a, we can define it *intrinsically* by derivations on the affine coordinate ring R of X.

If $a \in X$, then a derivation $D:R \longrightarrow \mathbb{C}$ *centered at a* means a $\mathbb{C}$-linear map such that

    i) $D(fg) = f(a) \cdot D(g) + g(a) \cdot D(f)$,

    ii) $D(\alpha) = 0$, all $\alpha \in \mathbb{C}$.

Clearly a derivation $D:R \longrightarrow \mathbb{C}$ at a is the same thing as a derivation D': $\mathbb{C}[X_1, \ldots, X_n] \longrightarrow \mathbb{C}$ at a such that $D'(f) = 0$, all $f \in \mathfrak{P}$. But a D' is determined by its values $\lambda_i = D(X_i)$ and conversely, given any $\lambda_1, \ldots, \lambda_n$ there is a D' with these values on the X_i. Since

$$D'f = \sum_{i=1}^{n} \frac{\partial f}{\partial X_i}(a) \cdot DX_i, \qquad \text{all } f \in \mathbb{C}[X],$$

the derivations D' which kill $\mathfrak{P}$ correspond to the $\lambda_1, \ldots, \lambda_n$ such that

$$\sum_{i=1}^{n} \frac{\partial f}{\partial X_i}(a) \cdot \lambda_i = 0, \qquad \text{all } f \in \mathfrak{P}.$$

This proves that

(1.9) $T_{X,a} \left(\begin{array}{c} \text{as vector space} \\ \text{with origin } a \end{array} \right) \cong \left\{ \begin{array}{c} \text{vector space of derivations} \\ D:R_X \longrightarrow \mathbb{C} \text{ centered at } a \end{array} \right\}$

We can make the definition more local too. First introduce:

(1.10) **Definition.** *If* $X = V(\mathfrak{P}) \subset \mathbb{C}^n$ *is an affine variety and* $x \in X$, *then define in any of 3 ways(!):*

$\mathcal{O}_{x,X} = $ *ring of rational functions* $\dfrac{f(X_1, \ldots, X_n)}{g(X_1, \ldots, X_n)}$ *where* $g(x) \neq 0$, *modulo those with* $f \in \mathfrak{P}$

$\qquad = $ *localization of* R_X *with respect to the multiplicative set of* $g, g(x) \neq 0$
$\qquad = $ *ring of germs of functions* $U \longrightarrow \mathbb{C}, U$ *a Zariski neighborhood of* x *in* X, *defined by rational functions* f/g, *where* $g(x) \neq 0$.

$\mathcal{O}_{x,X}$ is a local ring, since the set of functions f/g which are zero at X forms a maximal ideal and every f/g which is not zero at x is invertible, i.e., $g/f \in \mathcal{O}_{x,X}$. $\mathcal{O}_{x,X}$ is called the local ring of $x \in X$.
    The point is that every derivation $D:R_X \longrightarrow \mathbb{C}$ extends uniquely to the ring $\mathcal{O}_{x,X}$ by the rule $D(f/g) = (g(a)Df - f(a)Dg)/g(a)^2$, hence

$$T_{X,a} \cong \left\{ \begin{array}{c} \text{vector space of derivations} \\ D:\mathcal{O}_{x,X} \to \mathbb{C} \text{ centered at } a \end{array} \right\}.$$

Note that every function $f \in \mathcal{O}_{x,X}$ defines a linear map $df: T_{X,x} \longrightarrow \mathbb{C}$, its *differential* by the rule $df(D) = D(f)$. If $T_{X,x}$ is considered *externally* as a subspace of $\mathbb{C}^n$, then df is nothing but the linear term $\sum \frac{\partial f}{\partial X_i}(a) \cdot (X_i - a_i)$ in the Taylor expansion of f at a.

R_X and all the local rings $\mathcal{O}_{x,X}$ have the same quotient field, which we will denote $\mathbb{C}(X)$ and call the *function field* of X. An important fact is that the local rings $\mathcal{O}_{x,X}$ determine the affine ring R_X. In fact:

(1.11) **Proposition.** *Taking intersections in* $\mathbb{C}(X)$, *we have*:

$$R_X = \bigcap_{x \in X} \mathcal{O}_{x,X}.$$

Proof. The inclusion "$\subset$" is clear. Now say $f \in \mathcal{O}_{x,X}$ for all $x \in X$. Consider the ideal in $\mathbb{C}[X_1, \ldots, X_n]$ defined by:

$$\mathfrak{A} = \{ g \in \mathbb{C}[X_1, \ldots, X_n] \,|\, \text{if } \bar{g} = (g \bmod \mathfrak{P}) \in R_X, \quad \bar{g} \cdot f \in R_X \}.$$

Since $f \in \mathcal{O}_{x,X}$ we can write $f = h_x/g_x$, where $h_x, g_x \in \mathbb{C}[X_1, \ldots, X_n]$, $g_x(x) \neq 0$. Thus $g_x \in \mathfrak{A}$, hence $x \notin V(\mathfrak{A})$ for all $x \in X$. Moreover, $\mathfrak{A} \supset \mathfrak{P}$, so $V(\mathfrak{A}) \subset V(\mathfrak{P}) = X$. Therefore $V(\mathfrak{A}) = \phi$. But then by the Nullstellensatz $1 \in I(V(\mathfrak{A})) = \sqrt{\mathfrak{A}}$, hence $1 \in \mathfrak{A}$. By definition, this means $f \in R_x$.   QED

We next use the derivations of $\mathbb{C}(X)$ to prove:

(1.12) **Proposition.** $\exists$ *a non-empty Zariski open subset* $U \subset X$ *such that*:

$$\text{tr.d.}_{\mathbb{C}} \mathbb{C}(X) = \dim T_{X,a}, \text{ for all } a \in U.$$

Proof. It is well known that the transcendence degree of any separably generated field extension K/L is equal to the dimension of the K-vector space of derivations $D : K \longrightarrow K$ that kill L. In our case,

$$\text{tr.d.}_{\mathbb{C}} \mathbb{C}(X) = \dim_{\mathbb{C}(X)}(\text{derivations } D: \mathbb{C}(X) \longrightarrow \mathbb{C}(X) \text{ that kill } \mathbb{C})$$

$$= \dim_{\mathbb{C}(X)}(\text{derivations } D: R_X \longrightarrow \mathbb{C}(X) \text{ that kill } \mathbb{C})$$

$$= \dim_{\mathbb{C}(X)} \left(\begin{array}{l} \text{derivations } D: \mathbb{C}[X_1, \ldots, X_n] \longrightarrow \mathbb{C}(X) \\ \text{that kill } \mathbb{C} \text{ and } \mathfrak{P} \end{array} \right)$$

$$= \dim_{\mathbb{C}(X)} \left(\begin{array}{l} n\text{-tuples } \lambda_1, \ldots, \lambda_n \in \mathbb{C}(X) \text{ such that} \\ \sum \dfrac{\partial f_j}{\partial X_i} \cdot \lambda_i = 0 \text{ in } \mathbb{C}(X), \text{ all } j \end{array} \right)$$

$$= n - rk \left(\begin{array}{l} \text{the image in } \mathbb{C}(X) \text{ of the matrix} \\ (\partial f_i / \partial X_j)_{\substack{1 \leq i \leq l \\ 1 \leq j \leq n}} \end{array} \right)$$

Therefore it suffices to show

$$\left[rk \text{ in } \mathbb{C}(X) \text{ of } \left(\frac{\partial f_i}{\partial X_j} \right)_{\substack{1 \leq i \leq l \\ 1 \leq j \leq n}} \right] = \left[rk \frac{\partial f_i}{\partial X_j}(a)_{\substack{1 \leq i \leq l \\ 1 \leq j \leq n}} \right],$$

for all a in a Zariski open subset of X. But if r is the rank in $\mathbb{C}(X)$, then there is an invertible $l \times l$ matrix A over $\mathbb{C}(X)$ and an invertible $n \times n$ matrix B over $\mathbb{C}(X)$ such that:

$$A \cdot \left(\frac{\partial f_i}{\partial X_j} \bmod \mathfrak{P} \right) \cdot B = \left(\begin{array}{c|c} I_r & 0 \\ \hline 0 & 0 \end{array} \right)$$

Write $A = A_0/\alpha$, $B = B_0/\beta$ where A_0, B_0 are matrices over R_X and $\alpha, \beta \in R_X$. If U is the Zariski open set of X where $\det A_0 \cdot \det B_0 \cdot \alpha \cdot \beta \neq 0$, then for all $a \in U$,

$$\frac{A_0(a)}{\alpha(a)} \cdot \left(\frac{\partial f_i}{\partial X_j}(a) \right) \cdot \frac{B_0(a)}{\beta_0(a)} = \left(\begin{array}{c|c} I_r & 0 \\ \hline 0 & 0 \end{array} \right)$$

hence

$$rk \frac{\partial f_i}{\partial X_j}(a) = r. \quad QED$$

(1.13) **Definition.** $tr.d._{\mathbb{C}}\mathbb{C}(X)$, or equivalently $\min_{a \in X}(\dim T_{X,a})$, is called the dimension of X. A point $a \in X$ is called a smooth point or a singular point of X according as $\dim T_{X,a} = \dim X$ or $\dim T_{X,a} > \dim X$.

A simple example is given by the cuspidal cubic X defined by $X_1^2 = X_2^3$ in $\mathbb{C}^2$. This is the locus of points $(a^3, a^2), a \in \mathbb{C}$. Then $\dim X = 1$ and

i) $T_{X,(0,0)} = [\text{the whole plane } \mathbb{C}^2]$,

hence $(0,0)$ is a singular point,

ii) if $a \neq 0$, $T_{X,(a^3,a^2)} = [\text{the line defined by } 2(X_1 - a^3) = 3a(X_2 - a^2)]$

hence (a^3, a^2) is a smooth point.

(1.14) **Proposition.** *If X is a proper subvariety of Y, then $\dim X < \dim Y$.*

Proof. In fact, $X = V(\mathfrak{P})$, and $Y = V(\mathfrak{Q})$ where $\mathfrak{P} \supsetneq \mathfrak{Q}$ are 2 prime ideals. Let $R = \mathbb{C}[X]/\mathfrak{P}, S = \mathbb{C}[X]/\mathfrak{Q}$. Then $R \cong S/(\mathfrak{P}/\mathfrak{Q})$ so the proposition follows from:

(1.15) **Lemma.** *Let R be an integral domain over a field k, $\mathfrak{P} \subset R$ a prime. Then $tr.d._k R \geq tr.d._k R/\mathfrak{P}$, with equality only if $\mathfrak{P} = \{0\}$ or both sides are ∞. [By convention, $tr.d.\ R$ is the tr.d. of the quotient field of R.]*

Proof. Say $\mathfrak{P} \neq 0$, $tr.d._k R = n < \infty$. If the statement is false, there are n elements $x_1, \dots, x_n$ in R such that their images $\bar{x}_i$ in $R/\mathfrak{P}$ are algebraically independent. Let $0 \neq p \in \mathfrak{P}$. Then $p, x_1, \dots, x_n$ cannot be algebraically independent over k, so there is a polynomial $P(Y, X_1, \dots, X_n)$ over k such that $P(p, x_1, \dots, x_n) = 0$. Since R is an integral domain, we may assume P is irreducible. The polynomial P cannot equal $\alpha Y, \alpha \in k$, since $p \neq 0$. Therefore P is not even a multiple of Y. But then $P(0, \bar{x}_1, \dots, \bar{x}_n) = 0$ in $R/\mathfrak{P}$ is a nontrivial relation on the $\bar{x}_1, \dots, \bar{x}_n$. QED

As a consequence, every variety X has a stratification into *locally closed**
smooth sets:

a) Let $U = $ Zariski open set of smooth points
b) Decompose $X - U$ into components:

$$X - U = X_1^{(1)} \cup \dots \cup X_k^{(1)}.$$

Each $X_i^{(1)}$ is a variety of lower dimension than X. Let $U_i^{(1)}$ be the Zariski open set in $X_i^{(1)}$ of points smooth on $X_i^{(1)}$ (although singular on X).
c) Decompose $X - [U \cup U_1^{(1)} \cup \dots \cup U_k^{(1)}]$ into components:

$$X - [U \cup U_1^{(1)} \cup \dots \cup U_{(k)}^{(1)}] = X_1^{(2)} \cup \dots \cup X_l^{(2)}.$$

*Locally closed means open in its closure, or equivalently the intersection of an open and a closed set.

Each $X_i^{(2)}$ is a proper subvariety of an $X_j^{(1)}$ hencé has dimension at most $\dim X - 2$. Let $U_i^{(2)}$ be the Zariski open set in $X_i^{(2)}$ of points smooth on $X_i^{(2)}$. etc. Since dim is a positive integer, the process terminates and we have X written as a finite union of locally closed smooth varieties.

Another consequence is that every $(n-1)$-dimensional variety X in $\mathbb{C}^n$ is a hypersurface, i.e., $X = V(f), f$ irreducible. In fact, take any non-zero g that vanishes on X, and write $g = \Pi g_i, g_i$ irreducible. Then

$$X \subset \cup V(g_i),$$

so since X is irreducible, $X \subset V(g_i)$ for some i. But

$$\dim X \doteq n - 1 = \dim V(g_i),$$

so $X = V(g_i)$.

Our next step is to give a simple procedure for constructing a variety with a smooth point at the origin 0.

(1.16) Theorem. *Let $f_1, \ldots, f_r \in \mathbb{C}[X_1, \ldots, X_n]$ be polynomials with no constant term and independent linear terms. Then*

$$\mathfrak{P} = \{g \in \mathbb{C}[X_1, \ldots, X_n] \mid g = \sum \frac{h_i}{k} \cdot f_i, \quad h_i, k \in \mathbb{C}[X], k(0) \neq 0\}$$

is a prime ideal in $\mathbb{C}[X_1, \ldots, X_n], X = V(\mathfrak{P})$ has dimension $n - r$ and 0 is a smooth point of X. Moreover $V(f_1, \ldots, f_r) = X \cup Y$ where Y is a closed algebraic set with $0 \notin Y$.

Proof. The essential point is to relate the 3 rings:

$$\mathbb{C}[X_1, \ldots, X_n] \subset \mathbb{C}[X_1, \ldots, X_n]_{(X_1, \cdots, X_n)} \subset \mathbb{C}[[X_1, \ldots, X_n]]$$

$$\parallel \qquad\qquad\qquad \parallel \qquad\qquad\qquad\qquad \parallel$$

$$R_{\mathbb{A}^n} \qquad\qquad \mathcal{O}_{0,\mathbb{A}^n} \qquad\qquad \text{formal power series}$$

$$\begin{pmatrix} \text{local ring} \\ \text{of } 0 \in \mathbb{A}^n \end{pmatrix} \qquad \text{in } X_1, \ldots, X_n$$

Here the 2nd inclusion is a consequence of the expansion:

$$(1 - \sum X_i a_i)^{-1} = 1 + \sum_{n=1}^{\infty} (\sum X_i a_i)^n$$

$$\text{for any } a_i \in \mathbb{C}[X_1, \ldots, X_n].$$

It is easy to see that in each of these rings an arbitrary element f can be uniquely expanded

$$f = \sum_{|\alpha| < k} c_\alpha X^\alpha + \text{remainder } r, \quad c_\alpha \in \mathbb{C}$$

$$r \in \text{ideal generated by the monomials } X^\alpha, |\alpha| = k.$$

Therefore it makes sense to talk, for instance, of a function in any of these rings "vanishing to order k".

Now let $\mathfrak{P}'$ and $\mathfrak{P}''$ be the ideals generated by $f_1, \ldots, f_r$ in $\mathbb{C}[X]_{(X)}$ and $\mathbb{C}[[X]]$ respectively. Note that the ideal $\mathfrak{P}$ in the theorem is by definition $\mathfrak{P}' \cap \mathbb{C}[X]$, hence if $\mathfrak{P}'$ is prime, so is $\mathfrak{P}$. We claim also that

(1.17) $$\mathfrak{P}' = \mathfrak{P}'' \cap \mathbb{C}[X]_{(X)}$$

hence if $\mathfrak{P}''$ is prime, so are $\mathfrak{P}'$ and $\mathfrak{P}$.

But

$$\mathfrak{P}'' \cap \mathbb{C}[X]_{(X)} = \{ g \in \mathbb{C}[X]_{(X)} | g = \sum h_i f_i, h_i \in \mathbb{C}[[X]] \}$$

$$\subset \bigcap_{N=1}^{\infty} \left\{ g \in \mathbb{C}[X]_{(X)} \left| g = \sum h_i f_i + \left(\begin{matrix} \text{remainder vanishing} \\ \text{to order} \geq N \end{matrix} \right) \right. \right\}$$

$$\text{some } h_i \in \mathbb{C}[X] \text{ of degree} < N$$

$$= \bigcap_{N=1}^{\infty} \mathfrak{P}' + (X_1, \ldots, X_n)^N \cdot \mathbb{C}[X]_{(X)}.$$

But a theorem of Krull $(Z - S, \text{vol. 1, p. 217})$ states that if R is any local noetherian ring with maximal ideal $\mathfrak{M}$ and $\mathfrak{A}$ is any ideal, then "$\mathfrak{A}$ is closed in the $\mathfrak{M}$-adic topology", i.e.,

$$\mathfrak{A} = \bigcap_{N=1}^{\infty} \mathfrak{A} + \mathfrak{M}^N.$$

In our case, take $R = \mathbb{C}[X]_{(X)}$, $\mathfrak{A} = \mathfrak{P}'$, so that:

$$\mathfrak{P}' = \bigcap_{N=1}^{\infty} \mathfrak{P}' + (X_1, \ldots, X_n)^N \cdot \mathbb{C}[X]_{(X)}$$

which proves (1.17).

But now it is easy to show that $\mathfrak{P}''$ is prime using:

(1.18) **Formal implicit function theorem.** *Let*

$$f(X) = \sum_{i=1}^{n} a_i X_i + (\text{higher order terms}) \in \mathbb{C}[[X]]$$

and assume $a_1 \neq 0$. Then f can be factored:

$$f(X) = u(X_1, \ldots, X_n) \cdot (X_1 - g(X_2, \ldots, X_n))$$

$$\text{where } u(0) \neq 0, g(0) = 0,$$

and every other power series $k(X)$ can be written uniquely:

$$k(X) = a(X_1, \ldots, X_n) \cdot f(X_1, \ldots, X_n) + b(X_2, \ldots, X_n).$$

(1.18) is an easy exercise in power series so we omit its proof (alternately, it is a special case of the Weierstrass Preparation Theorem, $Z - S$, vol. II, p. 139). The last part may be restated as:

The residue ring $\mathbb{C}[[X_1, \ldots, X_n]]/(f)$ is

isomorphic to the subring $\mathbb{C}[[X_2, \ldots, X_n]]$.

Applying this inductively, we deduce:

(1.19) **Theorem.** *If* $f_i(X) = \sum a_{ij} X_j +$ (higher order terms), *and* $\det(a_{ij})_{\substack{1 \le i \le r \\ 1 \le j \le r}} \ne 0$, *then the residue ring* $\mathbb{C}[[X_1, \ldots, X_n]]/(f_1, \ldots, f_r)$ *is isomorphic to the subring* $\mathbb{C}[[X_{r+1}, \ldots, X_n]]$.

This applies to our given $f_1, \ldots, f_r$ after a suitable reordering of the variables. Since $\mathbb{C}[[X_{r+1}, \ldots, X_n]]$ is an integral domain, it follows that $\mathfrak{P}''$ is a prime ideal, hence so is $\mathfrak{P}$.

Next, to compute dim $V(\mathfrak{P})$, note first that

$$\mathbb{C}[X]/\mathfrak{P} \subset \mathbb{C}[[X]]/\mathfrak{P}'' \cong \mathbb{C}[[X_{r+1}, \ldots, X_n]],$$

hence $X_{r+1}, \ldots, X_n$ are algebraically independent in $\mathbb{C}[X]/\mathfrak{P}$. Therefore dim $V(\mathfrak{P}) \ge n - r$. On the other hand, the Zariski tangent space of $V(\mathfrak{P})$ at 0 is cut out by the linear terms of $f_1, \ldots, f_r$, so it follows that

$$\dim T_{V(\mathfrak{P}),0} = n - r.$$

This implies the other inequality dim $V(\mathfrak{P}) \le n - r$ and that x is a smooth point of $V(\mathfrak{P})$.

Finally, if $g_1, \ldots, g_s$ are generators of $\mathfrak{P}$, then for each g_i there is an h_i with $h_i(0) \ne 0$ and $h_i g_i \in (f_1, \ldots, f_s)$. If $h = \prod_{i=1}^{s} h_i$, it follows that $h\mathfrak{P} \subset (f_1, \ldots, f_r)$. Then:

$$V(\mathfrak{P}) \subset V(f_1, \ldots, f_r) \subset V(h\mathfrak{P}) = V(h) \cup V(\mathfrak{P}).$$

Since $0 \notin V(h)$, it follows that $V(f_1, \ldots, f_r)$ decomposes into $V(\mathfrak{P})$ plus a closed algebraic set Y with $0 \notin Y$.   QED

A Corollary of the theorem is that in fact every variety can be gotten by the procedure of (1.16):

(1.20) **Corollary.** *Let* $X = V(\mathfrak{P})$ *be a variety in* $\mathbb{C}^n$ *of dimension* $n - r$ *and let* $a = (a_1, \ldots, a_n)$ *be a smooth point on* X. *Since* dim $T_{a,X} = n - r$, *there are polynomials* $f_1, \ldots, f_r \in \mathfrak{P}$ *such that the linear terms of the* f_i *at a cut out* $T_{a,X}$. *Then*

(*)
$$\mathfrak{P} = \left\{ g \in \mathbb{C}[X] \,\middle|\, g = \sum \frac{h_i}{k} f_i, \, h_i, k \in \mathbb{C}[X], k(a) \ne 0 \right\}.$$

Proof. Let $\mathfrak{P}'$ be the ideal on the right in (*). By (1.16) applied around the origin a, $\mathfrak{P}'$ is prime $X' = V(\mathfrak{P}')$ is an $n - r$-dimensional variety. But clearly $\mathfrak{P} \supseteq \mathfrak{P}'$, hence $X \subseteq X'$. Since dim $X = $ dim X', this shows that $X = X'$, hence $\mathfrak{P} = \mathfrak{P}'$.   QED

§1B. Analytic Uniformization at Smooth Points, Examples of Topological Knottedness at Singular Points

We now turn to the topological and analytic structure of varieties. (1.20) can be used to *analytically uniformize* varieties near smooth points.

We first recall the definition of a complex manifold:

(1.21) **Definition.** *A complex manifold M of complex dimension n is a topological space together with a covering collection $\mathcal{U}$ of "charts", i.e., homeomorphisms*

$$\phi_\alpha : U_\alpha \xrightarrow{\;\approx\;} V_\alpha, \qquad \bigcup_\alpha U_\alpha = M$$
$$\underset{\text{open}}{\cap} \qquad \underset{\text{open}}{\cap}$$
$$M \qquad\qquad \mathbb{C}^n$$

such that $\phi_\beta \circ \phi_\alpha^{-1}$ is analytic where it is defined:

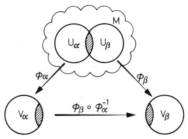

[Recall that a function F from an open set in $\mathbb{C}^n$ to an open set in $\mathbb{C}^n$ is *analytic* if each of its coordinates F_i is defined at each point of the domain by an absolutely convergent power series:

$$F_i(z_1,\dots,z_n) = \sum_{i_1,\cdots,i_n = 0}^{\infty} a_{i_1,\cdots,i_n} z_1^{i_1} \dots z_n^{i_n}$$

$$\text{or} = \sum_\alpha a_\alpha z^\alpha \text{ for short.}$$

Moreover, convergence is equivalent to

$$|a_{i_1,\cdots,i_n}| \le c_1 \cdot c_2^{i_1 + \cdots + i_n} \text{ for some } c_1, c_2 > 0.]$$

If one wants $\mathcal{U}$ to be really canonical at the expense of letting $\mathcal{U}$ be awfully big, it is customary to require $\mathcal{U}$ to be maximal with the above compatibility property.

In any case, any homeomorphism $\phi : U \xrightarrow{\;\approx\;} V, U \subset M, V \subset \mathbb{C}^n$, which is compatible with the charts ϕ_α is called a set of *local analytic coordinates* in U.

(1.22) **Definition.** *If M_1, M_2 are 2 complex manifolds, a holomorphic map $F : M_1 \longrightarrow M_2$ is a continuous map such that for all charts $\phi_\alpha : U_\alpha \xrightarrow{\;\sim\;} V_\alpha$, $U_\alpha \subset M_1, \psi_\beta : U_\beta \xrightarrow{\;\sim\;} V_\beta, U_\beta \subset M_2, \psi_\beta \circ F \circ \phi_\alpha^{-1}$ is analytic where is defined.*

I claim that the Zariski open set X_0 of smooth points on X has a natural structure of complex manifold. This follows from (1.16) and the convergent form of the implicit function theorem.

(1.23) **Analytic implicit function theorem.** *Let*

$$f(x) = \sum_{i=1}^{n} a_i X_i + (\text{higher order terms})$$

be a convergent power series with $a_1 \neq 0$. Then there is a unique convergent power series $g(X_2,\dots,X_n)$ and an $\varepsilon > 0$ such that if $|a_1|,\dots,|a_n| < \varepsilon$

$$f(a_1,\ldots,a_n) = 0 \Longleftrightarrow a_1 = g(a_2,\ldots,a_n).$$

One proof of this for instance is by showing that (1.18), the formal implicit function theorem, is true verbatim if f is assumed convergent and g and h are required to be convergent (cf. $Z-S$, vol. II, pp. 142–145). Then (1.23) is just a geometric translation. For another approach, cf. Narasimhan, [1], §1.3, pp. 142–145. Applying (1.23) inductively, we get:

(1.24) **2nd version.** *Let*

$$f_k(X) = \sum_{l=1}^{n} c_{kl} X_l + \text{(higher order terms)}, 1 \leq k \leq r$$

be a convergent power series, and assume $\det_{1 \leq k,l \leq r} a_{kl} \neq 0$. *Then there are unique convergent power series* $g_i(X_{r+1},\ldots,X_n), 1 \leq i \leq r$, *and an* $\varepsilon > 0$ *such that if* $|a_1|,\ldots,$ $|a_n| \leq \varepsilon$, *then*

$$\begin{pmatrix} f_i(a_1,\ldots,a_n) = 0 \\ 1 \leq i \leq r \end{pmatrix} \Longleftrightarrow \begin{pmatrix} a_i = g_i(a_{r+1},\ldots,a_n) \\ 1 \leq i \leq r \end{pmatrix}$$

(1.25) **3rd version.** *Let*

$$f_i(X_1,\ldots,X_r) = \sum a_{ij} X_j + \text{higher order terms}, 1 \leq i \leq r$$

be convergent power series and assume $\det a_{ij} \neq 0$. *Then there are unique convergent power series* $g_i(Y_1,\ldots,Y_r)$, $1 \leq i \leq r$ *and open sets* $U,V \subset \mathbb{C}^r$ *containing* $(0,\ldots,0)$ *such that*

$$y_i = f_i(x_1,\ldots,x_r) \text{ and } x_i = g_i(y_1,\ldots,y_r) \quad 1 \leq i \leq r$$

define inverse isomorphisms from U to V and from V to U.

Combining (1.20) and (1.24), we deduce:

(1.26) **Corollary.** *If* $a = (a_1,\ldots,a_n)$ *is a smooth point of an r-dimensional variety* $X \subset \mathbb{C}^n$, *let* $x_{i_1},\ldots,x_{i_r}$ *be any set of coordinates such that* $dx_{i_1},\ldots,dx_{i_r}$ *are independent linear functions on* $T_{a,X}$. *Then there is an* $\varepsilon > 0$ *and convergent power series* $g_j(x_{i_1},\ldots,x_{i_r})$, *for all* $j \in \{1,\ldots,n\} - \{i_1,\ldots,i_r\}$, *such that if* $|t_1|,\ldots,|t_n| < \varepsilon$

$$((a_1+t_1,\ldots,a_n+t_n) \in X) \Longleftrightarrow \begin{pmatrix} t_j = g_j(t_{i_1},\ldots,t_{i_r}), \\ \text{for all} \\ j \in \{1,\ldots,n\} - \{i_1,-,i_r\}. \end{pmatrix}$$

If we let $U_a = X \cap \{(a_1+t_1,\ldots,a_n+t_n, \text{ all } |t_i| < \varepsilon\}$ and V_a be the projection of U_a on the $(i_1,\ldots,i_r)$-coordinate plane $\mathbb{C}^r$, then we can cover the smooth points X_0 of X by the charts:

$$\phi_a : U_a \xrightarrow{\text{projection}} V_a$$
$$\cap \qquad\qquad \cap$$
$$X_0 \qquad\qquad \mathbb{C}^r$$

(1.26) shows that ϕ_a is a homeomorphism and that $\phi_b \circ \phi_a^{-1}$ is analytic wherever defined, i.e., X_0 is a complex manifold.

At singular points, varieties will not be submanifolds of $\mathbb{C}^n$ however and often are even topologically locally knotted. We would like to illustrate some of the ways this occurs. We take the case of a hypersurface $V(f) \subset \mathbb{C}^n$ on which the origin 0 is an isolated singular point. The best way to visualize the topology is to intersect $V(f)$ with a small sphere around 0. In fact, let

$$B_\varepsilon = \{(x_1, \ldots, x_n) \mid \sum |x_i|^2 \leq \varepsilon\}$$

Then $\partial B_\varepsilon \approx S^{2n-1}$ and it can be shown that if ε is small enough, $V(f) \cap \partial B_\varepsilon$ will be a real submanifold of ∂B_ε of dimension $2n-3$ and that $V(f) \cap B_\varepsilon$ will be homeomorphic to a cone over $V(f) \cap \partial B_\varepsilon$. (For these facts and much more on this beautiful link between topology and algebraic geometry, see Milnor [2]). To work out some examples, it helps to use stereographic projection to put coordinates on ∂B (take $\varepsilon = 1$ here):

Equations of Stereographic Projection in Real Coordinates

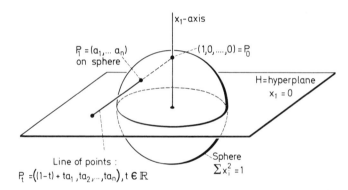

Then $P_t \in H \Longleftrightarrow t = \dfrac{1}{1 - a_1}$, in which case $P_t = \left(0, \dfrac{a_2}{1 - a_1}, \ldots, \dfrac{a_n}{1 - a_1}\right)$

Referring to the figure, we see that Stereographic projection:

$$(x, Y) \qquad \begin{array}{c} \partial B \subset \mathbb{C}^2 \\ \S \downarrow \\ R^3 \cup (\infty) \end{array}$$

is given in complex coordinates x, y by:

$$(x, y) \longrightarrow \left(\underbrace{\frac{\operatorname{Im} x}{1 - \operatorname{Re} x}}_{u}, \underbrace{\frac{\operatorname{Re} y}{1 - \operatorname{Re} x}}_{v}, \underbrace{\frac{\operatorname{Im} y}{1 - \operatorname{Re} x}}_{w}\right).$$

Example a. Let L_1 be the line $x = 0$ and L_2 the line $Y = 0$. Then

$$L_1 \cap \partial B = \text{the circle } u = 0, v^2 + w^2 = 1 \text{ in } \mathbb{R}^3 \cup (\infty)$$

$$L_2 \cap \partial B = \text{the circle } \{v = w = 0, u \in \mathbb{R}\} \text{ plus } \infty$$

Of course the inside $L_i \cap B$ is just a cone over $L_i \cap \partial B$. It follows that for the reducible variety $X = L_1 \cup L_2 = V(xy)$ with 2 smooth components through 0,

$$X \cap \partial B = (2 \text{ linked circles in the 3-sphere } \mathbb{R}^3 \cup (\infty))$$

and that $X \cap B$ is a cone over these 2 linked circles.

Example b. Let $X = V(x^2 - y^3)$. All points on X can be written uniquely as

$$\begin{aligned} x &= t^3 \\ y &= t^2 \end{aligned} \qquad t \in \mathbb{C}$$

and $(x, y) \in \partial B$ iff $|t|^6 + |t|^4 = 1$, i.e., $|t| = \lambda$ where λ is the unique positive solution of $\lambda^6 + \lambda^4 = 1$. Therefore

$$X \cap \partial B = \text{locus of points } (\lambda^3 e^{3i\theta}, \lambda^2 e^{2i\theta})_{0 \le \theta \le 2\pi}.$$

Under stereographic projection, it is easy to see that the torus:

$$T = \{|x| = \alpha, |y| = \beta, \alpha^2 + \beta^2 = 1\} \text{ in } \partial B$$

goes over to the torus

$$T' = \left\{ (u, v, w) \,\middle|\, u^2 + \left(\sqrt{v^2 + w^2} - \frac{1}{\beta} \right)^2 = \frac{\alpha^2}{\beta^2} \right\}$$

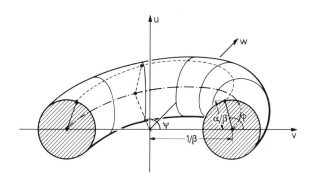

Cutaway view of T'

where the angular coordinates $\arg x, \arg y$ on T correspond to the angular coordinates ϕ, ψ on T'. Now $X \cap \partial B$ is a circle on T such that, as you go around once, $\arg x$ "goes around" three times and $\arg y$ "goes around" twice. On T', seen from above, this comes out as:

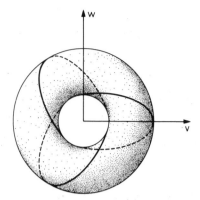

i.e., a trefoil knot.

It is not hard to see that by a homeomorphism of B leaving ∂B fixed, you can carry $X \cap B$ into a cone over $X \cap \partial B$.

Example c. Look at $X = V(2xy - z^2) \subset \mathbb{C}^3$. All points on X can be represented:

$$\begin{aligned} x &= s^2 \\ y &= t^2 \qquad\qquad s, t \in \mathbb{C} \\ z &= \sqrt{2}s \cdot t \end{aligned}$$

but this representation is not unique. In fact, (s,t) and (s',t') give the same point iff $(s,t) = (s',t')$ or $(-s', -t')$. Note that

$$|x|^2 + |y|^2 + |z|^2 = (|s|^2 + |t|^2)^2.$$

Therefore

$$X \cap \partial B \cong \left(\begin{array}{c} \text{unit sphere in } (s,t)\text{-space modulo} \\ (s,t) \sim (-s, -t) \end{array} \right)$$

$$= (\text{real projective 3-space, } \mathbb{R}\mathbb{P}^3).$$

Therefore a neighborhood $X \cap B$ of $0 \in X$ is homeomorphic to a cone over $\mathbb{R}\mathbb{P}^3$. Hence in this case X is not even a topological manifold near 0.

§1C. $\mathcal{O}_{x,X}$ a UFD when x Smooth; Divisor of Zeroes and Poles of Functions

We would like to conclude this chapter by looking again but with more algebraic tools at the local ring $\mathcal{O}_{x,X}$ in the case x is a smooth point on X. We saw in (1.16) the strength of power series expansions in $\mathbb{C}^n$. The fact that x is smooth on X implies that every function $f \in \mathcal{O}_{x,X}$ has a power series expansion *in local coordinates on X* as well. To be precise, say $X \subset \mathbb{C}^n$, $x = (0,\ldots,0)$, dim $X = r$ and $X = V(f_1,\ldots,f_{n-r})$ near 0 where df_i are independent at 0. Suppose $y_1,\ldots,y_r$ are any rational functions:

$$y_i = \frac{a_i(X_1,\ldots,X_n)}{b_i(X_1,\ldots,X_n)} \qquad a_i(0) = 0, \quad b_i(0) \neq 0$$

with $dy_1,\ldots,dy_r$ independent linear functions on $T_{x,X}$, i.e., $y_1,\ldots,y_r,f_1,\ldots,f_{n-r}$ all have independent linear terms at 0.

Look at the rings of (1.16) and their quotients by $f_1,\dots,f_{n-r}$:

$$
\begin{array}{ccc}
\mathcal{O}_{0,\mathbb{A}^n} & \longrightarrow & \mathbb{C}[[X_1,\dots,X_n]] \\
\downarrow & & \downarrow \\
\mathcal{O}_{0,\mathbb{A}^n}/(f_1,\dots,f_{n-r}) & \longrightarrow & \mathbb{C}[[X_1,\dots,X_n]]/(f_1,\dots,f_{n-r}) \\
\| \text{def} & & \mathfrak{s}\| \\
\mathcal{O}_{x,X} & & \mathbb{C}[[y_1,\dots,y_r]]
\end{array}
$$

(The isomorphism of $\mathbb{C}[[X]]/(f)$ and $\mathbb{C}[[y]]$ comes from (1.19)). Thus any $f \in \mathcal{O}_{x,X}$ has a natural expansion as a power series in $y_1,\dots,y_r$. The algebraic relationship of these two rings is given by:

(1.27) **Proposition.** *With the above notations, the quotients $\mathcal{O}_{x,X}/\mathfrak{M}_{x,X}^k$ and $\mathbb{C}[[y_1,\dots,y_r]]/(y_1,\dots,y_r)^k$ are isomorphic for all k, hence the formal completion:*

$$
\hat{\mathcal{O}}_{x,X} \underset{\text{def,}}{=} \varprojlim_k \mathcal{O}_{x,X}/\mathfrak{M}_{x,X}^k
$$

of $\mathcal{O}_{x,X}$ and the formal power series ring $\mathbb{C}[[y_1,\dots,y_r]]$ are isomorphic.

(Here $\mathfrak{M}_{x,X}$ = the ideal of $f \in \mathcal{O}_{x,X}$ such that $f(x) = 0$.) In fact, we saw in the proof of (1.16) that

$$
\mathcal{O}_{0,\mathbb{A}^n}/(\mathfrak{M}_{0,\mathbb{A}^n})^k \cong \mathbb{C}[[X_1,\dots,X_n]]/(X)^k
$$

$$
\cong \text{vector space of expressions } \sum_{|\alpha| < k} c_\alpha X^\alpha.
$$

Dividing both rings by the ideal generated by $(f_1,\dots,f_{n-r})$, they remain isomorphic. The isomorphism of $\mathcal{O}_{x,X}$ with $\mathbb{C}[[y_1,\dots,y_r]]$ follows by taking inverse limits, noting that $\mathbb{C}[[y_1,\dots,y_r]]$ is already complete. (For background on completions see $Z - S$, vol. II, p. 251 or $A - M.$, Ch. 10).

We use this to prove

(1.28) **Theorem.** *If $x \in X$ is a smooth point, then $\mathcal{O}_{x,X}$ is a UFD.*

Proof. We will not prove this from scratch, but rather reduce it to the elementary fact that formal power series rings are UFD's—see $Z - S$, vol. II, p. 148. The idea is to "pull down" the UFD property from the completion $\hat{\mathcal{O}}_{x,X}$, which we just saw was a formal power series ring, to $\mathcal{O}_{x,X}$ itself, just as we did for easier facts in the proof of (1.16). The first point in comparing these two rings is:

(1.29) **Lemma.** *If $a,b \in \mathcal{O}_{x,X}$ then a/b in $\hat{\mathcal{O}}_{x,X}$ implies a/b in $\mathcal{O}_{x,X}$.*

Proof. Here we use the result of Krull again:

$$
\bigcap_{k=1}^{\infty} [a \cdot \mathcal{O}_{x,X} + \mathfrak{M}_{x,X}^k] = a \cdot \mathcal{O}_{x,X}.
$$

If $a|b$ in $\hat{\mathcal{O}}_{x,X}$, then $b = a \cdot c$, some $c \in \hat{\mathcal{O}}_{x,X}$. Writing $c = \lim c_n$, where $c_n \in \mathcal{O}_{x,X}$ and $c - c_n \in \hat{\mathfrak{M}}_{x,X}^n$, then

$b = a \cdot c_n + a \cdot (c - c_n) \in a \cdot \mathcal{O}_{x,X} + $ (remainder vanishing to order n), i.e., $b \in$

$$\bigcap_{k=1}^{\infty} [a \cdot \mathcal{O}_{x,X} + m_{x,X}^k] = a \cdot \mathcal{O}_{x,X}.$$

Thus $a|b$ in $\mathcal{O}_{x,X}$.

Now since $\mathcal{O}_{x,X}$ is noetherian, the UFD property amounts to:

(*)   if $f|gh$, then $f = f_1 f_2$ where $f_1|g$ and $f_2|h$.

OK — start with $f, g, h \in \mathcal{O}_{x,X}$ such that $f|gh$. In $\hat{\mathcal{O}}_{x,X}$, let $d = g.c.d.(f,g)$ and let $f = d \cdot f'$, $g = d \cdot g'$. Write $f' = \lim f'_n$, $g' = \lim g'_n$ where $f'_n, g'_n \in \mathcal{O}_{x,X}$ and $f' - f'_n$, $g' - g'_n \in \hat{\mathfrak{M}}_{x,X}^n$. Then

$$f \cdot g'_n - g \cdot f'_n = f \cdot (g'_n - g') + g(f' - f'_n) \in f \cdot \hat{\mathfrak{M}}_{x,X}^n + g \cdot \hat{\mathfrak{M}}_{x,X}^n.$$

Therefore

$$f \cdot g'_n - g \cdot f'_n \in \mathcal{O}_{x,X} \cap [(f,g)\hat{\mathfrak{M}}_{x,X}^n \cdot \hat{\mathcal{O}}_{x,X}].$$

But if $\mathfrak{A} \subset \mathcal{O}_{x,X}$ is any ideal,

$$\mathcal{O}_{x,X} \cap \mathfrak{A} \cdot \hat{\mathcal{O}}_{x,X} \subset \bigcap_{k=1}^{\infty} (\mathfrak{A} + \mathfrak{M}_{x,X}^k) = \mathfrak{A}$$

by Krull's theorem, hence in particular

$$f \cdot g'_n - g \cdot f'_n = -f \cdot s_n + g \cdot r_n, \quad s_n, r_n \in \mathfrak{M}_{x,X}^n$$

i.e., (**) $f \cdot (g'_n + s_n) = g(f'_n + r_n)$

Therefore in $\hat{\mathcal{O}}_{x,X}$

$$f' \cdot (g'_n + s_n) = g'(f'_n + r_n)$$

and since f' and g' are relatively prime,

$$f'_n + r_n = h \cdot f', \qquad \text{some } h \in \hat{\mathcal{O}}_{x,X}.$$

Now choose n large enough so that $f' \notin \hat{\mathfrak{M}}_{x,X}^n$, i.e., $f'_n = f' \neq 0$ in $\mathcal{O}_{x,X}/\mathfrak{M}_{x,X}^n$. Then the h above cannot be in $\hat{\mathfrak{M}}_{x,X}$, so h must be a unit in $\hat{\mathcal{O}}_{x,X}$! Therefore

$$f'_n + r_n | f', \quad \text{and} \quad f' | f, \quad \text{hence} \quad f'_n + r_n | f \text{ in } \hat{\mathcal{O}}_{x,X}.$$

Applying (1.29), it follows that $f'_n + r_n | f$ in $\mathcal{O}_{x,X}$, i.e., $f = d_n \cdot (f'_n + r_n)$. Therefore by (**) $g = d_n \cdot (g'_n + s_n)$ and moreover $(f'_n + r_n)|(g'_n + s_n) \cdot h$. But in $\hat{\mathcal{O}}_{x,X}$, $f'_n + r_n$ and $g'_n + s_n$ differ from f' and g' only by units so they are relatively prime. Therefore in $\hat{\mathcal{O}}_{x,X}$, $f'_n + r_n | h$. By (1.29) again, $f'_n + r_n | h$ in $\mathcal{O}_{x,X}$. This proves (*) with $f_1 = d_n$, $f_2 = f'_n + r_n$.   QED

The geometric meaning of the UFD property is the following generalization of (1.16) and (1.20).

(1.30) **Corollary.** Let $X = V(\mathfrak{P})$ be an r-dimensional variety in $\mathbb{C}^n$ and let $x \in X$ be a smooth point. If $f(X)$ represents an irreducible element of $\mathcal{O}_{x,X}$, then

$$\mathfrak{P}' = \{g \in \mathbb{C}[X_1, \ldots, X_n] \mid kg \in \mathfrak{P} + (f), \text{ some } k \in \mathbb{C}[X], \text{ with } k(x) \neq 0\}$$

is a prime ideal, and $X' = V(\mathfrak{P}')$ is a subvariety of X containing x of dimension $r - 1$. Conversely any subvariety $X' \subset X$ containing x and of dimension $r - 1$ is obtained in this way from an irreducible element f of $\mathcal{O}_{x,X}$, characterized by the property: $I(X') \cdot \mathcal{O}_{x,X} = f \cdot \mathcal{O}_{x,X}$. Finally, for any $f \in R_X$, the components of $V(f)$ through x are all of dimension $r - 1^*$ and correspond in a one-one way as above with the irreducible factors of f in $\mathcal{O}_{x,X}$.

Proof. Let $\phi : \mathbb{C}[X_1, \ldots, X_n] \longrightarrow \mathcal{O}_{x,X}$ be the natural homomorphism. Then by definition, $\mathfrak{P}' = \phi^{-1}(f \cdot \mathcal{O}_{x,X})$, and since $\mathcal{O}_{x,X}$ is a UFD, the irreducible element f generates a prime ideal $f \cdot \mathcal{O}_{x,X}$. Therefore $\mathfrak{P}'$ is a prime ideal. Now $X' \subsetneqq X$, hence $\dim X' \leq r - 1$ by (1.14). On the other hand, $df = 0$ defines the Zariski tangent space $T_{x,X'}$, as a subspace of $T_{x,X}$, and, in fact, it is even true that for $y \in X'_0$, some Zariski open neighborhood of $x, df = 0$ defines $T_{y,X'}$ in $T_{\bar{y},X}$. [If $f_1, \ldots, f_i$ generate $\mathfrak{P}'$, then $k_i f_i = a_i p_i + b_i f$. Let $X'_0 = \{y \mid k_i(y) \neq 0, \text{ all } i\}$.] Therefore

$$\dim T_{y,X'} \geq \dim T_{y,X} - 1 \geq r - 1, \qquad \text{all } y \in X'_0.$$

Thus $\dim X' = r - 1$. Conversely, given $X' \subset X$ with $x \in X'$ and $\dim X' = r - 1$, look at the ideal $I(X') \cdot \mathcal{O}_{x,X}$. It is readily seen to be prime and hence factoring some element in it, it contains some non-trivial irreducible element $f(X_1, \ldots, X_n)/g(X_1, \ldots, X_n)$. Since $g(x) \neq 0$, g is a unit in $\mathcal{O}_{x,X}$ and we can assume this irreducible element is just f. But then the ideal $\mathfrak{P}'$ constructed from f satisfies $\mathfrak{P}' \subseteq I(X')$ [i.e., $\mathfrak{P} + (f) \subseteq I(X')$ and if $k \cdot g \in \mathfrak{P} + (f), k(x) \neq 0$, then $k \notin I(X')$, hence $g \in I(X')$ too]. Therefore $V(\mathfrak{P}') \supseteq X'$ and $\dim V(\mathfrak{P}') = r - 1 = \dim X'$ implies $X' = V(\mathfrak{P}')$. Therefore also $\mathfrak{P}' = I(X')$ and $f \cdot \mathcal{O}_{x,X} = \mathfrak{P}' \cdot \mathcal{O}_{x,X} = I(X') \cdot \mathcal{O}_{x,X}$. We leave the proof of the last assertion to the reader.   QED

For instance, if X is a curve and $x \in X$ is a smooth point, then the maximal ideal $\mathfrak{M}_{x,X}$ itself is principle and if f is a generator, then every other element $g \in \mathcal{O}_{x,X}$ equals $u \cdot f^n$, u a unit and $n \geq 0$.

(1.31) **Definition.** *An element $f \in \mathcal{O}_{x,X}$ such that $f \cdot \mathcal{O}_{x,X} = I(X') \cdot \mathcal{O}_{x,X}$ is called a local equation of X' at x.*

Now suppose the whole variety X is smooth. The situation would be very simple if the affine ring R_X were also a UFD: then every $f \in R_X$ would have an essentially unique decomposition:

$$f = (\text{unit}) \cdot \prod f_i^{r_i}, \qquad f_i \text{ irreducible}$$

which would correspond to the unique decomposition of algebraic sets:

$$V(f) = \cup Z_i, \qquad \dim Z_i = r - 1$$

via a bijection between equivalence classes of irreducible f's and subvarieties $Z \subset X$ of dimension $r - 1$. Unfortunately this does not usually happen and the situation is instead much more analogous to that in the rings of integers of algebraic number fields where the UFD property must be weakened to the more abstract

---

*We will see in (3.14) that this assertion is true at singular points $x \in X$ too.

decomposition of principal ideals into products of prime ideals. What we can do is this:

(1.32) **Definition.** *Let X be a smooth affine variety of dimension r.*

A) *Let* $\mathrm{Div}(X) = \left[\begin{array}{l}\text{free abelian group on the set of subvarieties} \\ Z \subset X \text{ of dimension } r - 1.\end{array}\right]$

This is called the group of divisors on X.

Now if $f \in R_X$, it follows from (1.30) that if

$$V(f) = Z_1 \cup \ldots \cup Z_k$$

is the decomposition of $V(f)$ into its irreducible components, then $\dim Z_i = r - 1$ for all i. We can then define an *order of vanishing* of f on each Z_i as follows:

B) *Choose any $x \in Z_i$ and any local equation $f_i \in \mathcal{O}_{x,X}$ of Z_i. Write:*

$$f = \frac{k}{g} \cdot f_i^{r_i} : k, g \in R_X, k, g \not\equiv 0 \text{ on } Z_i.$$

Set

$$r_i = \mathrm{ord}_{Z_i}(f).$$

It is easy to check that r_i is independent of x and f_i, and that $r_i > 0$.

C) *For all $f \in R_X, f \neq 0$, define the divisor (f) of f in $\mathrm{Div}(X)$ to be:*

$$(f) = \sum_{i=1}^{k} r_i \cdot Z_i$$

or

$$= \sum_{\dim Z = r - 1} (\mathrm{ord}_Z f) \cdot Z,$$

(where $\mathrm{ord}_Z f$ is defined to be 0 when Z is *not* a component of $V(f)$).
It is easy to check that

$$(f \cdot g) = (f) + (g).$$

Therefore we assign to every $f \in \mathbb{C}(X) - (0)$ a divisor of *zeroes and poles* in X by writing $f = g/h$, $g, h \in R_X$, and setting

D)

$$\mathrm{ord}_Z f = \mathrm{ord}_Z g - \mathrm{ord}_Z h$$

and

$$(f) = (g) - (h)$$
$$= \sum_Z (\mathrm{ord}_Z f) \cdot Z.$$

Z is called a *pole* of f if $\mathrm{ord}_Z f < 0$; a *zero* of f if $\mathrm{ord}_Z f > 0$. The subgroup of $\mathrm{Div}(X)$ of divisors (f) is called the *group of principal divisors*.

The following are immediate consequences of these fundamental definitions:

E) *R_X is characterized inside $\mathbb{C}(X)$ as the set of elements f with no poles, i.e., $\mathrm{ord}_Z f \geq 0$, all Z; $\mathcal{O}_{x,X}$ is similarly the set of f such that $\mathrm{ord}_Z f \geq 0$ for all Z through x.*

Proof. Suppose first that $f, g \in \mathcal{O}_{x,X} - (0)$ and $\mathrm{ord}_Z f \geq \mathrm{ord}_Z g$ for all Z through x. Then every irreducible element k of $\mathcal{O}_{x,X}$ divides f at least as often as it divides g.

Since $\mathcal{O}_{x,X}$ is a UFD, this means that $f/g \in \mathcal{O}_{x,X}$. This proves the 2nd part of E. Suppose next that $f, g \in R_X - (0)$ and $\mathrm{ord}_z f \geq \mathrm{ord}_z g$ for all Z. By the first argument, $f/g \in \mathcal{O}_{x,X}$ for all $x \in X$. Since by (1.11):

$$R_X = \bigcap_{x \in X} \mathcal{O}_{x,X},$$

it follows that $f/g \in R_X$.

F) *Finally, we may define* $\mathrm{Pic}(X)$ *to be the group* $\mathrm{Div}(X)/(\text{princ. divisors})$. *Then* R_X *itself is a UFD if and only if* $\mathrm{Pic}(X) = (0)$.
(Proof left to reader.)

Questions such as the structure and size of $\mathrm{Pic}(X)$ both in the algebraic number field case and in our geometric case have been a powerful magnet for research for 150 years. A good reference for some of the results in Serre [1].

Chapter 2. Projective Varieties

§ 2A. Their Definition, Extension of Concepts from Affine to Projective Case

Complex projective n-space can be defined in 3 ways:

i) as the set of $(n+1)$-tuples $a_0,\ldots,a_n$ of complex numbers, not all zero, modulo the equivalence relation

$$(a_0,\ldots,a_n) \sim (\lambda a_0,\ldots,\lambda a_n), \qquad \lambda \in \mathbb{C} - (0),$$

ii) as the set of 1-dimensional complex subspaces of the vector space $\mathbb{C}^{n+1}$,

iii) as the unit sphere $\{(a_0,\ldots,a_n) \mid \sum_{i=0}^{n} |a_i|^2 = 1\}$ in $\mathbb{C}^{n+1}$ modulo the equivalence relation:

$$(a_0,\ldots,a_n) \sim (e^{i\theta}a_0,\ldots,e^{i\theta}a_n), \quad 0 \le \theta \le 2\pi.$$

In general we use the 1st way. The resulting space will be denoted $\mathbb{P}^n$ and an $(n+1)$-tuple $a = (a_0,\ldots,a_n)$ defining a point P of $\mathbb{P}^n$ will be called a set of homogeneous coordinates of P. For simplicity we will often use the phrase: let $a = (a_0,\ldots,a_n)$ be a point of $\mathbb{P}^n$ although literally we mean the equivalence class of a is the point. $\mathbb{P}^n$ carries the quotient topology inherited from $\mathbb{C}^{n+1} - (0)$. This topology is clearly the same as the quotient topology of the unit sphere by the equivalence relation in (iii) so $\mathbb{P}^n$ is compact. We refer to this as the *classical topology on $\mathbb{P}^n$*.

$\mathbb{P}^n$ can also be viewed as $\mathbb{C}^n$ plus a "hyperplane at infinity". In fact, let

$$H_i = \text{the set of points } a \text{ in } \mathbb{P}^n \text{ such that } a_i = 0.$$

Then $\mathbb{P}^n - H_i$ is naturally isomorphic to $\mathbb{C}^n$ by the map:

$$(a_0,\ldots,a_n) \longrightarrow \left(\frac{a_0}{a_i},\ldots,\frac{a_{i-1}}{a_i},\frac{a_{i+1}}{a_i},\ldots,\frac{a_n}{a_j}\right)$$

We will call a_j/a_i, $j \in \{0,1,\ldots,\hat{i},\ldots,n\}$ the affine coordinates of a point a in the affine open set $\mathbb{P}^n - H_i$. Each H_i is itself naturally isomorphic to $\mathbb{P}^{n-1}$ (by simply disregarding the ith coordinate), so $\mathbb{P}^n$ can be regarded as the result of compactifying $\mathbb{C}^n$ by suitably glueing $\mathbb{P}^{n-1}$ at infinity. Also

$$\bigcap_{i=0}^{n} H_i = \phi$$

since no point a has all homogeneous coordinates a_j zero. Therefore

$$\bigcup_{i=0}^{n} (\mathbb{P}^n - H_i) = \mathbb{P}^n,$$

and we have a very explicit set of charts defining $\mathbb{P}^n$ as the union of $(n+1)$ copies of $\mathbb{C}^n$. Thus $\mathbb{P}^n$ is also in a natural way a complex analytic manifold.

Let $GL(n+1,\mathbb{C})$ be the group of $(n+1)\times(n+1)$ invertible complex matrices and let $PGL(n+1,\mathbb{C})$ be the quotient of $GL(n+1,\mathbb{C})$ by the diagonal matrices $\lambda\cdot I_{n+1}, \lambda\in\mathbb{C}-(0)$. Then $PGL(n+1,\mathbb{C})$ acts on $\mathbb{P}^n$ in an obvious way, and all our definitions will be invariant under the automorphisms of $\mathbb{P}^n$, known as *projective transformations*, gotten in this way. This shows that there is nothing sacred about the particular hyperplanes H_i. We may look at any hyperplane:

$$H = \text{locus of solutions of } \sum_{i=0}^{n} b_i X_i = 0, \quad (\text{not all } b_i = 0)$$

and we find that

$$\mathbb{P}^n - H \cong \mathbb{C}^n$$

via

$$(a_0,\ldots,a_n) \longrightarrow \left(\frac{a_0}{\Sigma b_i a_i}, \cdots\cdots, \frac{a_n}{\Sigma b_i a_i}\right)$$

↑
omit one of these for which the corresponding
b is $\neq 0$.

(2.1) Definition. *A closed algebraic set in $\mathbb{P}^n$ is a subset of the form*

$$V(f_1,\ldots,f_N) = \left\{ P\in\mathbb{CP}^n \middle| \begin{array}{l} \text{if } (a_0,\ldots,a_n) \text{ are homogeneous coordinates} \\ \text{of } P, \text{ then } f_1(a_0,\ldots,a_n) = \ldots = f_N(a_0,\ldots,a_n) = 0 \end{array} \right\}$$

where $f_1,\ldots,f_N$ are homogeneous polynomials in $\mathbb{C}[X_0,\ldots,X_n]$. (Homogeneity is necessary in order that the condition $f_i(a_0,\ldots,a_n) = 0$ be independent of the homogeneous coordinates chosen for the point P).

If $\mathfrak{A}$ is the ideal generated by $f_1,\ldots,f_N$, then we denote $V(f_1,\ldots,f_N)$ also by $V(\mathfrak{A})$. Note that $\mathfrak{A}$ is a homogeneous ideal, i.e., if a polynomial g is in $\mathfrak{A}$, and $g = \Sigma g_i, g_i$ homogeneous of degree i, then each g_i is in $\mathfrak{A}$. Conversely any homogeneous ideal $\mathfrak{A}$ has a finite set of homogeneous generators. Moreover if $\mathfrak{A}^\lambda = \{f(\lambda x_0,\ldots,\lambda x_n)|f\in\mathfrak{A}\}$ then it is easy to see that an ideal $\mathfrak{A}\subset\mathbb{C}[X_0,\ldots,X_n]$ is homogeneous if $\mathfrak{A} = \mathfrak{A}^\lambda$ for some infinite set of λ's in $\mathbb{C}-(0)$*. As a consequence, given a homogeneous $\mathfrak{A}$, if

$$\sqrt{\mathfrak{A}} = \mathfrak{P}_1 \cap \ldots \cap \mathfrak{P}_s,$$

then

(2.2) $\qquad \mathfrak{A}$ homogeneous $\Longrightarrow \begin{array}{l} \mathfrak{A} = \mathfrak{A}^\lambda, \text{ all } \lambda \\ \Longrightarrow \sqrt{\mathfrak{A}} = \sqrt{\mathfrak{A}^\lambda}, \text{ all } \lambda \text{ and } \forall\, i, \exists\, j \text{ such} \\ \text{that } \mathfrak{P}_i = \mathfrak{P}_j^\lambda \text{ for an infinite set} \end{array}$

$$\lambda = \lambda_1, \lambda_2, \ldots$$

*Say $\mathfrak{A} = \mathfrak{A}^{\lambda_i}, i = 0,1,2,\ldots$. Now if $f = \sum_{k=0}^{n} f_k$ is a polynomial in $\mathfrak{A}$ and f_k is homogeneous of degree k, then

$$f^{\lambda_i} = \Sigma\lambda_i^k f_k \in \mathfrak{A}, \qquad i = 0,1,2,\ldots.$$

But $\det(\lambda_i^k)_{0\leq i,k\leq n} \neq 0$, so each $f_k\in\mathfrak{A}$ (cf. $Z-S$, vol. II, p. 155).

$$\Longrightarrow \sqrt{\mathfrak{A}} \text{ homogeneous and } \mathfrak{P}_i^{\lambda_j \lambda_1^{-1}} = \mathfrak{P}_i, \text{ all } j,$$
hence all the $\mathfrak{P}_i$ are homogeneous too.

The operation V has the same properties as in the affine case:

a) $V\left(\sum_{i\in S}\mathfrak{A}_i\right) = \bigcap_{i\in S} V(\mathfrak{A}_i)$

b) $V(\mathfrak{A}_1 \cap \mathfrak{A}_2) = V(\mathfrak{A}_1) \cup V(\mathfrak{A}_2)$

c) If $\mathfrak{A}_1 \subseteq \mathfrak{A}_2$, then $V(\mathfrak{A}_1) \supseteq V(\mathfrak{A}_2)$

d) $V(\mathfrak{A}) = V(\sqrt{\mathfrak{A}})$, and if $\sqrt{\mathfrak{A}} = \mathfrak{P}_1 \cap \ldots \cap \mathfrak{P}_s$,
   then $V(\mathfrak{A}) = V(\mathfrak{P}_1) \cup \ldots \cup (\mathfrak{P}_s)$.

In particular, we can define the sets $V(\mathfrak{A})$ to be the closed sets of the *Zariski topology* and we call $V(\mathfrak{P})$ a *variety* if $\mathfrak{P}$ is a homogeneous prime ideal.

We now have two topologies on $\mathbb{P}^n$ which are completely different: a) the classical or usual topology whose open sets in the affine pieces are unions of the "polycylinders" $|y_1 - a_1| < \varepsilon, \ldots, |y_n - a_n| < \varepsilon$; and b) the Zariski topology with vastly bigger open sets $\{x | f(x) \neq 0\}$.

One feature of the operation V on projective space not analogous to V for affine space is:

e) all homogeneous ideals are contained in the one homogeneous maximal ideal $(X_0, X_1, \ldots, X_n)$ and $V(X_0, \ldots, X_n) = \phi$.

The following Nullstellensatz can be proved exactly as in §1 by constructing a generic point:

f) If $\mathfrak{P}$ is a homogeneous prime ideal, $\mathfrak{P} \neq (X_0, \ldots, X_n)$ and f is a homogeneous polynomial such that $f \notin \mathfrak{P}$, then there is an $x \in V(\mathfrak{P})$ such that $f(x) \neq 0$.

(2.3) Corollary. *For all homogeneous $\mathfrak{A}$ and all homogeneous f of degree $d \geq 1$, $f \in \sqrt{\mathfrak{A}} \Longleftrightarrow f$ vanishes identically on $V(\mathfrak{A})$.*

(2.4) Corollary $V(\mathfrak{A}) = \phi \Longleftrightarrow \sqrt{\mathfrak{A}} = (X_0, \ldots, X_n)$.

From the unique factorization property of $\mathbb{C}[X_0, \ldots, X_n]$, it follows that minimal homogeneous prime ideals are principal, hence

g) The maximal subvarieties $X \subsetneqq \mathbb{P}^n$ are the hypersurfaces $V(f)$, f an irreducible homogeneous polynomial.

If $X \subset \mathbb{P}^n$ is a subvariety corresponding to the homogeneous prime ideal $\mathfrak{P}$, $\mathbb{C}[X_0, \ldots, X_n]/\mathfrak{P}$ is called the *homogeneous coordinate ring* of X.

We want to investigate the connection between affine and projective varieties:

(2.5) Proposition. *Identify $\mathbb{C}^n$ with $\mathbb{P}^n - H_0$ as above and let $Y_i = \dfrac{X_i}{X_0}, 1 \leq i \leq n$ be affine coordinates in $\mathbb{C}^n$. This identification is a homeomorphism of $\mathbb{C}^n$ in its Zariski topology with $\mathbb{P}^n - H_0$ in the restriction of the Zariski topology. Moreover*

a) *if $\mathfrak{P} \subset \mathbb{C}[X_0, \ldots, X_n]$ is a homogeneous prime ideal defining the projective variety X, then the affine variety $X - X \cap H_0$ is $V(\mathfrak{P}')$, where*

$$\mathfrak{P}' = \{f(1, Y_1, \ldots, Y_n) | f \in \mathfrak{P}\}.$$

b) *if* $\mathfrak{P} \subset \mathbb{C}[Y_1, \ldots, Y_n]$ *is a prime ideal defining the affine variety* X, *then its Zariski closure* $\bar{X}$ *in* $\mathbb{P}^n$ *is* $V(\mathfrak{P}')$, *where* $\mathfrak{P}'$ *is the ideal generated by the homogenizations*

$$\tilde{f}(X) = \Sigma a_{i_1, \ldots, i_n} X_0^{(d - \Sigma i_k)} X_1^{i_1} \ldots X_n^{i_n} \text{ of the polynomials}$$

$$f(Y) = \Sigma a_{i_1, \ldots, i_n} Y_1^{i_1} \ldots Y_n^{i_n} \text{ in } \mathfrak{P} \; (d = \text{degree of } f).$$

$\left[\vphantom{\dfrac{X_1}{X_0}}\right.$ *Note* $\tilde{f}$ *can also be defined by*

$$\tilde{f}(X_0, \ldots, X_n) = X_0^d \, f\left(\frac{X_1}{X_0}, \ldots, \frac{X_n}{X_0}\right). \Big]$$

Proof. The quickest way to see the 1st part is to note that the open sets:

$$\mathbb{C}_f^n = \{a \in \mathbb{C}^n | f(a) \neq 0\}, \qquad f \in \mathbb{C}[Y_1, \ldots, Y_n]$$

are a *basis* for the Zariski topology of $\mathbb{C}^n$, and that the open sets

$$\mathbb{P}_f^n = \{a \in \mathbb{P}^n | f(a) \neq 0\}, \qquad f \in \mathbb{C}[X_0, \ldots, X_n]$$
$$f \text{ homogeneous}$$

are a basis in $\mathbb{P}^n$. But if $f \in \mathbb{C}[X]$ is homogeneous and $g(Y) = f(1, Y_1, \ldots, Y_n)$, then $\mathbb{P}_f^n - (H_0 \cap \mathbb{P}_f^n) = \mathbb{C}_g^n$; and if $f \in \mathbb{C}[Y]$ and $\tilde{f}(X)$ is its homogenization, then $\mathbb{P}_{\tilde{f}}^n - (H_0 \cap \mathbb{P}_{\tilde{f}}^n) = \mathbb{C}_f^n$. (a) and (b) are then almost tautologies.   QED

(2.6) *Example.* A good example to test your visualization of the line at ∞ is given by the plane conic

$$\Sigma = V(XY - Z^2) \subset \mathbb{P}^2.$$

In the affine $\mathbb{P}^2 - (X = 0)$, use affine coordinates $U = Y/X, V = Z/X$ and Σ becomes the parabola $U - V^2 = 0$. However, in the affine $\mathbb{P}^2 - (Z = 0)$, if we use affine coordinates $S = Y/Z, T = Y/Z$, then Σ becomes the hyperbola $ST = 1$. Thus the parabola and the hyperbola are projectively the same although the parabola has only one "point at ∞", $\Sigma \cap (X = 0) = (0, 1, 0)$, while the hyperbola has 2: $\Sigma \cap (Z = 0) = \{(0, 1, 0), (1, 0, 0)\}$

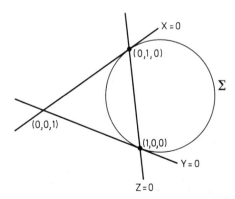

[As for the ellipse, it only looks different from the hyperbola over the reals and it has two conjugate imaginary points at ∞. Thus the projective approach to

Euclidean geometry is based on the observation that among all real quadratic polynomials $f(x, y) = 0$ defining ellipses, those defining circles are characterized by the fact that the corresponding complex curve in $\mathbb{P}^2$ passes through the famous "circular points at ∞" $(1, i, 0)$, $(1, -i, 0)$.]

As a consequence of (2.5), every projective variety has a Zariski open covering by affine varieties and hence we may extend the concepts of local rings, the function field, tangent space, dimension, singular and smooth points more or less automatically to the projective case.

Let $X = V(\mathfrak{P}) \subset \mathbb{P}^n$ be a projective variety.

(2.7) Definition.

a) *If $X_0 = X - X \cap H$ is an affine piece of X, $H = V(\Sigma a_i X_i)$, then*
$$R_{X_0} = \begin{bmatrix} ring \ of \ rational \ functions \ f(X_0, \dots, X_n)/(\Sigma a_i X_i)^d \ where \ f \ is \ homo- \\ geneous \ of \ degree \ d, \ modulo \ the \ ideal \ of \ f/(\Sigma a_i X_i)^{d}\text{'s} \ with \ f \in \mathfrak{P}. \end{bmatrix}$$

b) *For all $\bar{x} \in X$, you can then define in any of three ways (!) a ring:*
$$\mathcal{O}_{x, X} = \begin{bmatrix} ring \ of \ rational \ functions \ \dfrac{f(X_0, \dots, X_n)}{g(X_0, \dots, X_n)}, f, g \ homogeneous \ of \ the \\ same \ degree, \ g(x) \neq 0, \ modulo \ the \ ideal \ of \ f/g\text{'s}, \ f \in \mathfrak{P} \end{bmatrix}$$
$$= \begin{bmatrix} ring \ of \ germs \ of \ functions \ from \ Zariski\text{-}open \ neighborhoods \ of \ x \\ on \ X \ to \ \mathbb{C} \ defined \ by \ rational \ functions \ f/g \end{bmatrix}$$
$$= \begin{bmatrix} the \ local \ ring \ \mathcal{O}_{x, X - X \cap H} \ of \ x \ on \ the \ affine \ varieties \ X - X \cap H \ for \\ any \ hyperplane \ H \ such \ that \ x \notin H. \end{bmatrix}$$

c) *Define a field:*
$$\mathbb{C}(X) = \begin{bmatrix} ring \ of \ rational \ functions \ \dfrac{f(X_0, \dots, X_n)}{g(X_0, \dots, X_n)}, f, g \ homogeneous \ of \ the \\ same \ degree, \ g \notin \mathfrak{P}, \ modulo \ the \ ideal \ of \ f/g\text{'s}, \ f \in \mathfrak{P} \end{bmatrix}$$
$$= [quotient \ field \ of \ \mathcal{O}_{x, X}, \ any \ x \in X]$$
$$= [function \ field \ as \ in \ \S 1 \ of \ any \ of \ the \ affine \ varieties \ X - X \cap H.]$$
We call this the function field of X.

d) *Define vector spaces:*
$$T_{x, X} = [derivations \ D : \mathcal{O}_{x, X} \to \mathbb{C} \ centered \ at \ x, \ i.e., \ D(fg) = f(x)Dg + g(x)Df]$$
$$= \begin{bmatrix} tangent \ space \ as \ in \ \S 1 \ to \ X - X \cap H \ at \ x, \ for \ any \ hyperplane \ such \\ that \ x \notin H. \end{bmatrix}$$

e) *Define* $\dim X = \operatorname{tr.d.}_{\mathbb{C}} \mathbb{C}(X) = \min_{x \in X} \dim T_{x, X}.$

Note that if $X \not\subset$ (hyperplane $X_0 = 0$), then the homogeneous coordinate ring $\mathbb{C}[X_0, \dots, X_n]/\mathfrak{P}$ of $X = V(\mathfrak{P})$ has quotient field $\mathbb{C}(X)(X_0)$, hence $\dim X = \operatorname{tr.d.}\mathbb{C}[X_0, \dots, X_n]/\mathfrak{P} - 1.$

f) *Define x smooth (resp. singular) iff $\dim T_{x, X} = \dim X$ (resp. $\dim T_{x, X} > \dim X$).*

As in §1, X is the disjoint union of "strata" each of which is a Zariski-locally closed subset $Y \subset X$ of points smooth on $\bar{Y}$ and where $k = \dim \bar{Y} < \dim X$ for all but the one open stratum consisting of the smooth points on X itself. And moreover, in the classical topology induced from $\mathbb{P}^n$, the whole of X is a compact topological space and the strata are analytic manifolds of various dimensions k.

We can illustrate all this by quadrics $X = V(F) \subset \mathbb{P}^n$, F homogeneous of degree 2. By the standard theory of quadratic forms, in suitable coordinates

$$F(X) = \sum_{i=0}^{r} X_i^2, \quad r = \text{``rank'' of } F.$$

In the affine space $\mathbb{P}^n - V(X_k)$, X is given by the affine equations:

$$1 + \sum_{i=0}^{r} \left(\frac{X_i}{X_k}\right)^2 = 0 \qquad \text{if} \quad 0 \leq k \leq r$$

or

$$\sum_{\substack{i=0 \\ i \neq k}}^{r} \left(\frac{X_i}{X_k}\right)^2 = 0 \qquad \text{if} \quad r < k \leq n.$$

The 1st equation defines a non-singular affine hypersurface, but the 2nd equation vanishes with all its first derivatives on $\dfrac{X_o}{X_k} = \ldots = \dfrac{X_r}{X_k} = 0$, hence is singular at these points. In other words, if $r = n$, X is non-singular; but if $r < n$, $S = V(X_0, \ldots, X_r)$ is the set of singular points on X. In this 2nd case, X has two strata: an $(n-1)$-dimensional open piece $X - S$ and an $(n-r-1)$-dimensional linear space S.

   In case the whole variety X is smooth, the theory of divisors also extends more or less automatically.

(2.8) **Definition.**
   a) *Let* $\mathrm{Div}(X) = $ *free abelian group on subvarieties* $Z \subset X$ *of codimension 1,*
   b) *If* $f \in \mathbb{C}(X) - (0)$ *and* $Z \subset X$ *has codimension 1, choose* $x \in Z$, *and* $f_Z \in \mathcal{O}_{x,X}$ *a local equation of* Z. *Then*

$$g \cdot f = k \cdot f_Z^r, \qquad \begin{array}{l} \text{some } r \in \mathbb{Z}, \quad g, k \in \mathcal{O}_{x,X} \\ \text{such that} \qquad g, k \not\equiv 0 \text{ on } Z \end{array}$$

Then r *is equal to* $\mathrm{ord}_Z f$ *as defined in §1 for any affine piece* $X - X \cap H$ *of* X *which contains* x; *and if* x_1, x_2 *are any two points of* Z, *the* r's *which result can be compared by choosing* a hyperplane* H *such that* $x_1, x_2 \notin H$ *and looking at the affine* $X - X \cap H$. *Thus* $r = \mathrm{ord}_Z f$ *is independent of the choice of* x *and* f_Z.
   c) *If* $f \in \mathbb{C}(X) - (0)$, *set*

$$(f) = \sum_{\mathrm{codim}\, Z = 1} (\mathrm{ord}_Z f) \cdot Z \in \mathrm{Div}(X).$$

   d) *As in the affine case, we get the important group*

$$\mathrm{Pic}(X) = \mathrm{Div}(X) / \{(f) | f \in \mathbb{C}(X) - (0)\}$$

and as well as the characterization of $\mathcal{O}_{x,X}$ *as the ring of* $f \in \mathbb{C}(X)$ *with no "poles" through* x *(i.e., no* Z *with* $\mathrm{ord}_Z f < 0$).
   A good example of these ideas in $X = \mathbb{P}^n$ itself. A divisor D on $\mathbb{P}^n$ can be written:

$$D = \sum n_i \cdot V(f_i)$$

where $f_i \in \mathbb{C}[X_0, \ldots, X_n]$ is an irreducible homogeneous polynomial. If $g \in \mathbb{C}(\mathbb{P}^n) -$

----

*Note that for any finite set of points $x_1, \ldots, X_m \in \mathbb{P}^n$, there is always a hyperplane H such that $x_i \notin H$, $1 \leq i \leq m$.

(0), write $g = g_1/g_2$, $g_i \in \mathbb{C}[X_0,\ldots,X_n]$ homogeneous, relatively prime and of the same degree. Factoring g_1 and g_2, we get

$$g = \prod f_i^{n_i}, \quad n_i \in \mathbb{Z}$$
$$f_i \text{ irreducible, distinct, homogeneous}$$
$$\text{polynomials}$$

hence:

$$(g) = \sum n_i V(f_i).$$

Note that in this case

$$0 = \deg g = \sum n_i \cdot \deg f_i.$$

Conversely, for any D, if $\sum n_i \cdot \deg f_i = 0$, then $\prod f_i^{n_i} \in \mathbb{C}(\mathbb{P}^n)$ and $D = (\prod f_i^{n_i})$. Therefore if we define

$$\deg D = \sum n_i \cdot \deg f_i,$$

we get an exact sequence:

$$0 \longrightarrow \mathbb{C}(\mathbb{P}^n)^*/\mathbb{C}^* \longrightarrow \mathrm{Div}(\mathbb{P}^n) \xrightarrow{\deg} \mathbb{Z} \longrightarrow 0$$

(here $\mathbb{C}(\mathbb{P}^n)^* = $ multiplicative group $\mathbb{C}(\mathbb{P}^n) - (0)$, $\mathbb{C}^* = $ subgroup $\mathbb{C} - (0)$), hence $\mathrm{Pic}(\mathbb{P}^n) \cong \mathbb{Z}$.

§2B. Products, Segre Embedding, Correspondences

The next step in the theory is the study of the product spaces $\mathbb{C}^n \times \mathbb{C}^m$, $\mathbb{C}^n \times \mathbb{P}^m$ and $\mathbb{P}^n \times \mathbb{P}^m$. To fix notation, let $\{X_i\}$ and $\{Y_j\}$ be coordinates in the 1st and 2nd factors of the product. We introduce a Zariski topology in these products by considering the zeroes of polynomials $f(X;Y)$ in both sets of variables. To be precise:

(2.9) Definition.
a) *in $\mathbb{C}^n \times \mathbb{C}^m$, closed sets are* $V(f_1,\ldots,f_N), f_i \in \mathbb{C}[X_1,\ldots,X_n,Y_1,\ldots,Y_m]$
b) *in $\mathbb{C}^n \times \mathbb{P}^m$, closed sets are* $V(f_1,\ldots,f_N), f_i \in \mathbb{C}[X_1,\ldots,X_n,Y_0,Y_1,\ldots,Y_m]$
where f_i is homogeneous in the Y's, i.e.,

$$f_i = \sum_{\substack{b_0 + \cdots + b_m = d \\ \text{any finite} \\ \text{set of } a's}} (\text{coeff.}) X_1^{a_1} \ldots X_n^{a_n} Y_0^{b_0} \ldots Y_m^{b_m}, \ d = degree \ in \ Y.$$

c) *in $\mathbb{P}^n \times \mathbb{P}^m$, closed sets are* $V(f_1,\ldots,f_N)$, $f_i \in \mathbb{C}[X_0,\ldots,X_n,Y_0,\ldots,Y_m]$ *where f_i's are homogeneous separately in the X's and Y's, i.e.,*

$$f_i = \sum_{\substack{a_0 + \cdots + a_n = d \\ b_0 + \cdots + b_n = e}} (\text{coeff.}) X_0^{a_0} \ldots X_n^{a_n} Y_0^{b_0} \ldots Y_m^{b_m}, \ (d,e) = bi\text{-}degree.$$

Note that in case (a), we have simply re-defined the Zariski topology on $\mathbb{C}^{n+m}$: there is nothing new here. In case (b) the important thing to note is that we have glued together in the Zariski topology subsets homeomorphic to case (a):

(2.10) **Proposition.**

$$\mathbb{C}^n \times \mathbb{P}^m \supset \begin{bmatrix} \mathbb{C}^n \times (\mathbb{P}^m - H_i) \; in \\ its \; induced \; topology \end{bmatrix} \underset{homeo.}{\approx} \mathbb{C}^{n+m}.$$

Proof. A basis for open sets on the left are the sets $f\left(X_1, \ldots, X_n; \dfrac{Y_0}{Y_i}, \ldots, \dfrac{Y_i}{Y_i}, \ldots, \right.$

$\left. \dfrac{Y_m}{Y_i}\right) \neq 0$, f homogeneous in the 2nd set of variables. A basis on the right are the

sets $g\left(X_1, \ldots, X_n; \dfrac{Y_0}{Y_i}, \ldots, \dfrac{Y_i}{Y_i}, \ldots, \dfrac{Y_m}{Y_0}\right) \neq 0$, g any polynomial. These are the same
sets if g is gotten from f by setting the appropriate variable equal to 1, and f is
gotten from g by homogenizing in the 2nd set of variables.

In case (c), similarly, the space is also a glued together version of $\mathbb{C}^{n+m}$'s:

(2.11) **Proposition.**

$$\mathbb{P}^n \times \mathbb{P}^m \supset \begin{bmatrix} (\mathbb{P}^n - H_i) \times (\mathbb{P}^m - H_j) \\ in \; its \; induced \; topology \end{bmatrix} \underset{homeo,}{\approx} \mathbb{C}^{n+m}.$$

(Proof similar to (2.10)).

Fortunately, the whole theory of projective product varieties is reduced to
the theory of plain projective varieties because of the remarkable:

(2.12) **Proposition.** *Let* $(Z_{ij})_{\substack{0 \leq i \leq n \\ 0 \leq j \leq m}}$ *be homogeneous coordinates in* $\mathbb{P}^{nm+n+m}$. *Let*
$\mathfrak{A}$ *be the homogeneous ideal generated by the quadratic polynomials* $Z_{ij}Z_{kl} - Z_{il}Z_{kj}$.
Then the map

$$s : \mathbb{P}^n \times \mathbb{P}^m \longrightarrow V(\mathfrak{A})$$

$$(x, y) \longrightarrow The \; point \; Z_{ij} = x_i y_j$$

is a homeomorphism, and $V(\mathfrak{A})$ *is a projective variety.**

Proof. To prove injectivity of s, suppose $s(x, y) = s(x', y')$. Then $x_i y_j = \lambda x_i' y_j'$
for some $\lambda \neq 0$, and all i, j. Now for some i_0 and j_0, $x_{i_0} \neq 0$ and $y_{j_0} \neq 0$. Therefore
$x_{i_0}' y_{j_0}' \neq 0$, hence $x_{i_0}' \neq 0$ and $y_{j_0}' \neq 0$. If $\mu = x_{i_0}/x_{i_0}'$, $v = y_{j_0}/y_{j_0}'$, then $\lambda = \mu v$ and

$$x_i y_{j_0} = \lambda x_i' y_{j_0}' = \mu v x_i' y_{j_0}' = \mu x_i' y_{j_0}$$
$$hence \; x_i = \mu x_i', \; all \; i$$
$$hence \; x = x'.$$

Similarly $y = y'$. To prove surjectivity of s suppose $(z_{ij})_{\substack{0 \leq i \leq n \\ 0 \leq j \leq n}}$ satisfy $z_{ij}z_{kl} = z_{il}z_{kj}$
and are not all zero. Say $z_{i_0, j_0} \neq 0$. Set $x_i = z_{ij_0}/z_{i_0, j_0}$, and $y_j = z_{i_0 j}/z_{i_0 j_0}$. Then

$$(z_{i_0 j_0})^2 x_i y_j = z_{ij_0} z_{i_0 j} = z_{ij} z_{i_0 j_0}$$

i.e., $s(x, y)$ is a point with homogeneous coordinates equivalent to z_{ij}. To prove

---

*In fact $\mathfrak{A}$ is a prime ideal, but we omit this because we do not need it.

the map is a homeomorphism, first note that the topology of $\mathbb{P}^n \times \mathbb{P}^m$ has a basis consisting of the sets $f(x,y) \neq 0$, f homogeneous of bi-degree (d,d), i.e., of the same degree in the X's and the Y's. In fact, we know a basis is given by $f \neq 0$, f of bi-degree (d,e). Say $d \geq e$. Then

$$\{(x,y)|f(x,y) \neq 0\} = \bigcup_{j=0}^{m} \{(x,y)|(Y_j^{d-e}f)(x,y) \neq 0\}$$

and $Y_j^{d-e}f$ has bi-degree (d,d). Secondly, any polynomial $f(X,Y)$ of bi-degree (d,d) can be written as a $F(\ldots, X_i Y_j, \ldots)$ where F is homogeneous of degree d. Therefore the open set $f \neq 0$ in $\mathbb{P}^n \times \mathbb{P}^m$ is $V(\mathfrak{A}) \cap (\text{open set } F \neq 0)$, and the map is a homeomorphism. Finally, to prove $V(\mathfrak{A})$ is a variety, it suffices to prove $V(\mathfrak{A})$ is irreducible in the Zariski topology. More generally any product of varieties is irreducible, hence is a variety in virtue of:

(2.13) **Lemma.** *Let X and Y be irreducible topological spaces and put any topology on $X \times Y$ inducing the given topology on the images $\{x\} \times Y$ of Y and $X \times \{y\}$ of X. Then $X \times Y$ is irreducible.*

Proof. Say $X \times Y = S \cup T$. For all $x \in X$, $\{x\} \times Y = S \cap (\{x\} \times Y) \cup T \cap (\{x\} \times Y)$. Since Y is irreducible, for all x, $\{x\} \times Y$ is contained in either S or T. Let $s_y : X \to X \times Y$ be the map $x \mapsto (x,y)$. Then

$$\bigcap_y s_y^{-1}(S) = \{x|(x,y)\in S, \text{ all } y\}$$
$$= \{x|\{x\} \times Y \subset S\}.$$

Call this closed set S'. Similarly let

$$T' = \bigcap_y s_y^{-1}(T) = \{x|\{x\} \times Y \subset T\}.$$

Then what we showed first can be expressed $S' \cup T' = X$. Since X is irreducible, $X = S'$ or $X = T'$. Therefore $X \times Y = S$ or $= T$.   QED

The mapping of (2.12) is known as the *Segre embedding*. From now on, when we talk of $\mathbb{P}^n \times \mathbb{P}^m$, we regard it as a projective variety by the above embedding. Three more basic but very easy facts about $\mathbb{P}^n \times \mathbb{P}^m$ are these:

(2.14) **Proposition.**
   a) *if $x \in \mathbb{P}^n$ and $X_1, \ldots, X_n$ are affine coordinates around x*
        $y \in \mathbb{P}^m$ and $Y_1, \ldots, Y_m$ are affine coordinates around y

then

$$\mathcal{O}_{(x,y),\mathbb{P}^n \times \mathbb{P}^m} = \left[\text{ring of germs of functions} \frac{f(X,Y)}{g(X,Y)}, \text{ where } g(x,y) \neq 0, \text{ from a} \atop \text{neighborhood of } (x,y) \text{ to } \mathbb{C}, \right]$$

   b) *if $x \in \mathbb{P}^m$, $y \in \mathbb{P}^m$, then the tangent space $T_{(x,y),\mathbb{P}^n \times \mathbb{P}^m}$ is canonically isomorphic to $T_{x,\mathbb{P}^n} \times T_{y,\mathbb{P}^m}$. In particular, the image is a non-singular variety of dimension $n + m$.*

   c) *s is a bi-holomorphic map of the complex manifold $\mathbb{P}^n \times \mathbb{P}^m$ with the image of s.*

   We omit proofs.

The case $n = m = 1$ is very nice. Here we have a homeomorphism of $\mathbb{P}^1 \times \mathbb{P}^1$ with the quadric $V(X_{00}X_{11} - X_{01}X_{10})$ in $\mathbb{P}^3$. This quadratic form has maximal rank, so in fact the image is non-singular as stated in (b). As we have seen, all non-singular quadrics are projectively equivalent and hence they are all isomorphic to $\mathbb{P}^1 \times \mathbb{P}^1$. The product structure on a non-singular quadric in $\mathbb{P}^3$ has a very simple projective meaning: in fact, the images of curves $\{x\} \times \mathbb{P}^1$ and $\mathbb{P}^1 \times \{y\}$ are easily checked to be straight lines, and conversely all straight lines on the quadric are images of curves of one of these two types. These are the two "rulings" on a quadric which are easily visualized in the case of the hyperbolic paraboloid $z = xy$:

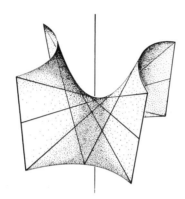

We are now ready to introduce the very basic definition:

(2.15) **Definition.** *Let $X \subset \mathbb{P}^n$ and $Y \subset \mathbb{P}^m$ be two varieties. A correspondence Z from X to Y is a relation given by a closed algebraic subset $Z \subset X \times Y$. Z is said to be a rational map if Z is irreducible and there is a Zariski open set $X_0 \subset X$ such that each $x \in X_0$ is related by Z to one and only one point of Y. Z is said to be a birational map if $Z \subset X \times Y$ and $Z^{-1} \subset Y \times X$ ($Z^{-1} = \{(y,x) | (x,y) \in Z\}$) are both rational maps.*

(2.16) ⇄ A "rational map from X to Y" restricts to a single-valued set-theoretic map from a Zariski open $X_0 \subset X$ to Y, but is not 'a map at all from the whole set X to Y: it can be "many-valued" on proper subvarieties of X.

(2.17) **Example.** The elements of the function field $\mathbb{C}(X)$ define rational maps from X to $\mathbb{P}^1$. In fact, suppose $X = V(\mathfrak{P})$ where $\mathfrak{P} \subset \mathbb{C}[X_0,\ldots,X_n]$ and $f,g \in \mathbb{C}[X_0,\ldots,X_n]$ are homogeneous of the same degree where $g \notin \mathfrak{P}$. Let Y_0, Y_1 be homogeneous coordinates in $\mathbb{P}^1$. Then consider the closed algebraic set

$$Z = V(Y_1 g - Y_0 f) \subset X \times \mathbb{P}^1.$$

If X_0 is the non-empty open set in X where $g \neq 0$ or $f \neq 0$, then if $x \in X_0$:

$$(x,y) \in Z \text{ iff } \frac{y_1}{y_0} = \frac{f(x)}{g(x)} \qquad \text{(both sides possibly } \infty\text{)}$$

i.e., restricted to X_0, Z is the graph of the map $x \to f(x)/g(x) \in \mathbb{P}^1$. Moreover

$Z \cap (X_0 \times \mathbb{P}^1)$ is homeomorphic to X_0 via projection onto its 1st component. This is because $p_1 : Z \cap (X_0 \times \mathbb{P}^1) \to X_0$ is continuous, bijective and closed, (we will prove that p_1 is closed shortly in (2.23)). Therefore $Z \cap (X_0 \times \mathbb{P}^1)$ is irreducible, and if we decompose Z into irreducibles, we get:

$$Z = Z^* \cup Y_1 \cup \ldots \cup Y_k$$
$$\text{where } Y_i \subset \left(\begin{array}{c} \text{proper subvariety} \\ \text{of } X \end{array} \right) \times \mathbb{P}^1$$

$$\text{but } p_1 : Z^* \to X \text{ is surjective by (2.23).}$$

Thus Z^* is the Zariski-closure of the graph of the map $x \to f(x)/g(x)$ and is a rational map from X to $\mathbb{P}^1$. We will see in Ch. 3 that in this way, $\mathbb{C}(X)$ is *isomorphic* to the set of rational maps from X to $\mathbb{P}^1$ (excluding the constant map with value ∞). If X is smooth, we can say a lot more about Z^*. We have defined the divisor (f) of f: write it

$$(f) = \underset{\substack{\text{divisor} \\ \text{of} \\ \text{"zeroes"} \\ \text{of } f}}{(f)_0} - \underset{\substack{\text{divisor} \\ \text{of} \\ \text{"poles"} \\ \text{of } f}}{(f)_\infty} \text{'(coefficients of } (f)_0 \text{ and } (f)_\infty \text{ positive).}$$

Then if $(f)_0 = \Sigma n_i D_i$, $(f)_\infty = \Sigma m_i E_i$, we can prove

(2.18) Proposition.
i) $\cup D_i = \{ x \in X \,|\, (x,0) \in Z^* \}$
ii) $\cup E_i = \{ x \in X \,|\, (x,\infty) \in Z^* \}$.

Proof. In fact, for all $x \in X$, using the fact that $\mathcal{O}_{x,X}$ is a UFD, write $f = a_x/b_x$, $a_x, b_x \in \mathcal{O}_{x,X}$ being relatively prime. Write $a_x = A_x/C_x$, $b_x = B_x/C_x$ where A_x, B_x, $C_x \in \mathbb{C}[X_0,\ldots,X_n]$ are homogeneous of the same degree and $C_x(x) \neq 0$. Then $B_x Y_1 - A_x Y_0$ vanishes on $Z \cap (X_0 \times \mathbb{P}^1)$, hence it vanishes on Z^* From this it follows that

$$x \notin \cup E_i \Longrightarrow b_x \text{ unit} \Longrightarrow B_x(x) \neq 0 \Longrightarrow (x,\infty) \notin Z^*.$$

To show that other implication, note first

$$\begin{array}{cccc} x \in \cup E_i, & \Longrightarrow b_x(x) = 0 & \Longrightarrow B_x(x) = 0 & \text{the only point } (x,\lambda) \\ \text{but } x \notin \cup D_i & \text{but } a_x(x) \neq 0 & \text{but } A_x(x) \neq 0 & \text{at which } B_x Y_1 - A_x Y_0 \\ & & & \text{vanishes is } (x,\infty). \end{array}$$

Since $Z^* \to X$ is surjective, this means that (x,∞) must be on Z^*. Thus

$$\cup E_i - \cup D_i \subset \{ x \in X \,|\, (x,\infty) \in Z^* \} \subset \cup E_i.$$

Since $\cup E_i$ is a Zariski closure of $\cup E_i - \cup D_i$, equality (ii) follows. (i) is similar.
                                                                                                         QED

While we are discussing this, here are two further results which we can prove by applying some basic theorems that will come soon:

(2.19) Proposition.

   i) *If* $\cup E_i = \phi$, *i.e., f has no poles, then* $f \in \mathbb{C}$.

   ii) $\cup D_i \cap \cup E_i = \{x \in X \mid \{x\} \times \mathbb{P}^1 \subset Z^*\}$
$$= \{x \in X \mid Z^* \text{ has more than one "value" at } x\}.$$

Proof. (i) follows from the fact that the projection $p_2(Z^*)$ on $\mathbb{P}^1$ is Zariski-closed (cf. (2.23)), hence if $\infty \notin p_2(Z^*)$, then $p_2(Z^*) =$ one point α, hence $f \equiv \alpha$. (ii) follows from the fact that the set of images $Z^*[x] \underset{\text{def,}}{=} \{\alpha \in \mathbb{P}^1 \mid (x, \alpha) \in Z^*\}$ is

connected (cf. (3.24)) hence is either one point or the whole line $\mathbb{P}^1$

(2.20) *Example.* Let $\mathbb{P}^2_X$ and $\mathbb{P}^2_Y$ be 2 planes with coordinates (X_0, X_1, X_2) and (Y_0, Y_1, Y_2) respectively. Consider the bijective map

$$f : \mathbb{P}^2_X - \left(\begin{array}{c}\text{triangle} \\ X_0 X_1 X_2 = 0\end{array}\right) \xrightarrow{\;\;\approx\;\;} \mathbb{P}^2_Y - \left(\begin{array}{c}\text{triangle} \\ Y_0 Y_1 Y_2 = 0\end{array}\right)$$

$$(x_0, x_1, x_2) \xrightarrow{\hspace{3cm}} \frac{1}{x_0}, \frac{1}{x_1}, \frac{1}{x_2}.$$

This extends to a remarkable birational correspondence from $\mathbb{P}^2_X$ to $\mathbb{P}^2_Y$. In fact, define

$$Z = V(X_0 Y_0 - X_1 Y_1, X_1 Y_1 - X_2 Y_2) \subset \mathbb{P}^2_X \times \mathbb{P}^2_Y.$$

Suppose $(x, y) \in Z$. Then $x_0 y_0 = x_1 y_1 = x_2 y_2 = \lambda$ say. There are three cases—

   (1) *if* $x_0 x_1 x_2 \neq 0$: then since some $y_i \neq 0$, it follows that $\lambda \neq 0$ and $y_i = \lambda/x_i$ all i, i.e., y equals $f(x)$.

   (2) *if* $x_0 = 0$, *but* $x_1 x_2 \neq 0$: then $\lambda = 0$, so $y_1 = y_2 = 0$. i.e., $y =$ the point $(1, 0, 0)$.

   (3) *if* $x_0 = x_1 = 0$, but $x_2 \neq 0$, i.e., $x = (0, 0, 1)$. Then $\lambda = 0$ and hence $y_2 = 0$ but there is no restriction on y_0, y_1.

Thus Z defines a correspondence that

   (1) is bijective on $\mathbb{P}^2_X - (X_0 X_1 X_2 = 0)$

   (2) "blows down" each line $X_i = 0$ (excluding the two points where it meets some $X_j = 0, j \neq i$) to one point.

   (3) "blows up" each point P_i with jth coordinate δ_{ij} to a whole line.

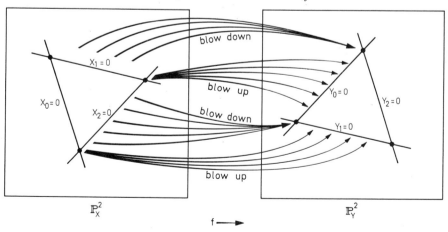

(2.21) *Example.* Let $\mathbb{P}^n_X$ and $\mathbb{P}^{n-1}_Y$ be spaces with coordinates $(X_0,\ldots,X_n),(Y_0,\ldots,Y_{n-1})$. Consider the so-called *projection*:

$$p:\mathbb{P}^n_X - (0,\ldots,0,1)\longrightarrow \mathbb{P}^{n-1}_Y$$
$$(x_0,\ldots,x_n)\longmapsto (x_0,\ldots,x_{n-1})$$

$P_n = (0,\ldots,0,1)$ is called the *center* of the projection. p is surjective and for all $x\in\mathbb{P}^n_X$, $x\neq P_n$, $p^{-1}(p(x))$ is the line joining P_n and x. Now p also extends to a rational correspondence from the whole variety $\mathbb{P}^n_X$ to $\mathbb{P}^{N-1}_Y$: define

$$Z = V(\ldots,Y_iX_j - Y_jX_i,\ldots)_{0\leq i,j\leq n-1}\subset \mathbb{P}^n_X \times \mathbb{P}^{n-1}_Y.$$

One sees immediately that Z is the union of the graph of p and the pairs (P_n,y), any $y\in\mathbb{P}^{n-1}_Y$. In other words the center P_n is "blown up" by Z to the whole $\mathbb{P}^{n-1}_Y$. Moreover Z is irreducible, hence is a variety itself: let $E = \{P_n\} \times \mathbb{P}^{n-1}_Y \subset Z$. Then by (2.23), $Z - E$ is homeomorphic to $\mathbb{P}^n_X -\{P_n\}$ hence is irreducible. But for all $y\in\mathbb{P}^{n-1}_Y$, let l_y be the line $p^{-1}(y)$ plus P_n. Then $(l_y - \{P_n\}) \times \{y\}\subset Z - E$. Therefore $l_y \times \{y\} \subset$ (Zariski closure of $Z - E$). In particular, $(P_n,y)\in\overline{Z-E}$, hence $E\subset \overline{Z-E}$, i.e., Z is irreducible. Z is interesting not only as a correspondence but as a variety in its own right. Z looks like $\mathbb{P}^n_X$ except that P_n has been replaced by a copy of $\mathbb{P}^{n-1}$. Z is called the blowing up of $\mathbb{P}^n$ at the point P_n—we will study it more thoroughly in Chapter 5. Blowings up are an essential tool in the problem of *resolution of singularities:* given a projective variety X, find a birational correspondence Z between X and a smooth projective variety X'. When dim $X = 1$, we will solve this twice: in Ch. 6 and in Ch. 7.

These examples illustrate the ways in which rational maps can be many-valued at special points. It is also important to pin down the set of points where a rational map behaves as nicely as possible.

(2.22) **Definition.** *Let $Z\subset X \times Y$ be a rational map from X to Y. Let $x\in X$ and let $Z[x] = \{y\in Y\,|\,(x,y)\in Z\}$. Then Z is regular at x if first of all Z is single-valued, i.e., $Z[x] = $ (a single point y). And secondly, if $(X_1,\ldots,X_n)$, resp. $(Y_1,\ldots,Y_m)$ are affine coordinates around x, resp. y, then in some Zariski neighborhood U of x, Z is the graph of a map:*

$$U \longrightarrow Y$$

given by

$$Y_i = \frac{a_i(X_1,\ldots,X_n)}{b_i(X_1,\ldots,X_n)}$$

(a_i,b_i polynomials such that b_i is nowhere zero on U). Z is called regular if it is regular at all points x.

By definition, $X_{\text{reg}} = \{x\in X\,|\,Z$ regular at $x\}$ is Zariski-open. We shall see later that it is always non-empty.

At regular points x, a rational map Z behaves just like maps to which we are accustomed, e.g., C^∞ maps of differentiable manifolds or analytic maps of analytic manifolds. Thus Z induces a "pull-back" homomorphism on the local rings:

$$Z^*:\mathcal{O}_{y,Y} \longrightarrow \mathcal{O}_{x,X}, \qquad y = Z[x].$$

To see this, note that for any correspondence Z, and $z = (x,y) \in Z$, we get homomorphisms:

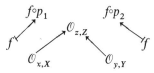

Then the regularity of Z means that $Y_i \circ p_2 = (a_i/b_i) \circ p_1$ on Z, and since every function in $\mathcal{O}_{z,Z}$ is of the form $c(X,Y)/d(X,Y)$ where $d(x,y) \neq 0$, this means that on Z

$$\frac{c(X,Y)}{d(X,Y)} = \frac{c\left(X, \dfrac{a(X)}{b(X)}\right)}{d\left(X, \dfrac{a(X)}{b(X)}\right)} = \frac{e(X)}{f(X)}, \qquad f(X) \neq 0,$$

hence $\mathcal{O}_{x,X} \longrightarrow \mathcal{O}_{z,Z}$ is *surjective*. This means it is also injective because $p_1(Z)$ is dense in X, hence $f \circ p_1 \equiv 0 \Longrightarrow f \equiv 0$. Thus $\mathcal{O}_{x,X} \longrightarrow \mathcal{O}_{z,Z}$ is an isomorphism, and $f \longrightarrow f \circ p_2$ induces $Z^* : \mathcal{O}_{y,Y} \longrightarrow \mathcal{O}_{z,Z} \cong \mathcal{O}_{x,X}$.

Having Z^* on local rings, we get other "functorial" properties for Z, e.g.,

(1) $dZ : T_{x,X} \longrightarrow T_{y,Y}$: namely, if $D : \mathcal{O}_{x,X} \longrightarrow \mathbb{C}$ is a derivation, then $f \longrightarrow D$ (Z^*f) is a derivation from $\mathcal{O}_{y,Y}$ to $\mathbb{C}$ which we call $dZ(D)$;

(2) assuming $p_2(Z)$ dense in Y, so that Z^* is an injective map from $\mathcal{O}_{y,Y}$ to $\mathcal{O}_{x,X}$, then Z^* induces an injective map $Z^* : \mathbb{C}(Y) \longrightarrow \mathbb{C}(X)$.

(3) Suppose X and Y are smooth and Z is everywhere regular. Then we may use Z^* to define a pull-back $Z^{-1} : \mathrm{Div}(Y) \longrightarrow \mathrm{Div}(X)$ too. Take a divisor D on Y. For all $x \in X$, let $f \in \mathcal{O}_{Z(x),Y}$ be a local equation of D at $Z(x)$. Define $Z^{-1}(D)$ to be the unique divisor with local equation $Z^*(f)$ at x.

Assuming (2.27) from the next section, it follows that the composition of 2 regular correspondences is again a regular correspondence. Therefore the set of all projective varieties and regular correspondences forms a *category*. One of the surprising things about classical geometry, however, is how infrequently this category is used. The point is that working projectively, the set of rational maps falls in your lap and the regular ones seemed classically to be merely a peculiar special type. It is only in the context of schemes that regular maps or morphisms begin to seem the more basic notion.

§2C. Elimination Theory, Noether's Normalization Lemma, Density of Zariski-Open Sets

The next result is central in all discussions of projective varieties: it is the algebraic counterpart of the compactness of $\mathbb{P}^n$ in the classical topology.

(2.23) Main Theorem of elimination theory. *The projection*

$$p_2 : \mathbb{P}^n \times \mathbb{P}^m \longrightarrow \mathbb{P}^m$$

is closed, i.e., if $Z \subset \mathbb{P}^n \times \mathbb{P}^m$ is a closed algebraic set, then so is $p_2(Z)$.

Proof. Since the result is local on the image, the theorem implies and is implied by the local version:

(2.24) $$p_2 : \mathbb{P}^n \times \mathbb{C}^m \longrightarrow \mathbb{C}^m \text{ closed.}$$

We will prove this form. Let $V(f_1, \ldots, f_N)$ be a closed algebraic set in $\mathbb{P}^n \times \mathbb{C}^n$ where $f_i \in \mathbb{C}[X_0, \ldots, X_n, Y_1, \ldots, Y_m]$, and f_i is homogeneous in the X's of degree d_i. For all $y \in \mathbb{C}^m$:

$$y \notin p_2(V(f_1, \ldots, f_N)) \iff f_i(X, y), \ 1 \leq i \leq N, \text{ have no common zeroes in } \mathbb{P}^n$$

$$\iff \exists d \geq 1 \text{ such that}$$
$$(X_0, \ldots, X_n)^d \subset (f_1(X, y), \ldots, f_N(X, y))$$

by Cor. 2 of the Nullstellensatz. Therefore, it suffices to prove that for each $d \geq 1$,

$$\{y \in \mathbb{C}^m \,|\, (X_0, \ldots, X_n)^d \subset (f_1(X, y), \ldots, f_N(X, y))\}$$

is a Zariski open set in $\mathbb{C}^m$. Let V_k be the vector space of homogeneous polynomials of degree k in $X_0, \ldots, X_n$, and consider for each $y \in \mathbb{C}^m$, the linear mapping:

$$V_{d-d_1} \oplus \cdots \cdots \oplus V_{d-d_N} \xrightarrow{\ T^{(d)}(y)\ } V_d$$

$$(g_1, \cdots \cdots \cdots, g_n) \longrightarrow \Sigma f_i(X, y) \cdot g_i(X).$$

In terms of some fixed basis of these two vector spaces, $T_d(y)$ can be written out as a $n_d \times m_d$-matrix $(T_{ij}^{(d)}(y))$, whose entries are polynomials in the y's. It follows that

$$(X_0, \ldots, X_n)^d \subset (f_1(X, y), \ldots, f_N(X, y)) \iff T^{(d)}(y) \text{ is surjective}$$

$$\iff \exists m_d \times m_d \text{ minor of } T_{ij}^{(d)}(y)$$
$$\text{whose determinant is} \neq 0.$$

This proves that the set of $y \in \mathbb{C}^m$ with this property is indeed Zariski-open.   QED

The classical approach to (2.23) is based on resultants and has the advantage that it eliminates the non-constructive step in the above proof where we take the union over d. We sketch this approach. First I recall the properties of the resultant:

$$\text{Let } f(X) = a_0 X^n + a_1 X^{n-1} + \ldots + a_n$$
$$g(X) = b_0 X^m + b_1 X^{m-1} + \ldots + b_m$$

be general polynomials. Their resultant is a polynomial

$$R = R(a_0, \ldots, a_n; b_0, \ldots, b_m)$$

with integer coefficients such that:

(2.25)

a) $R(a, b) \equiv A(a, b, X) f(X) + B(a, b, X) \cdot g(X)$ for suitable integral polynomials A and B,

b) if k is an algebraically closed field and $\bar{f}(X) = \Sigma \bar{a}_i X^{n-i}, \bar{g}(X) = \Sigma \bar{b}_j X^{m-j}$ are particular polynomials over k, then $R(\bar{a}, \bar{b}) = 0$ if and only if $\bar{a}_0 = \bar{b}_0 = 0$ or $\bar{f}$ and $\bar{g}$ have a common root, i.e., if and only if the homogeneous polynomials $\Sigma \bar{a}_i X^{n-i} Y^i$, $\Sigma \bar{b}_i X^{m-i} Y^i$ have a common root in $k^2 - (0,0)$. (See Lang [1], p. 135; van der Waerden [2], vol. I, §5.8).

The first step in the classical proof is to reduce the problem in some way to the case $n = 1$, i.e., the closedness of $\mathbb{P}^1 \times \mathbb{C}^m \to \mathbb{C}^m$. (One method is to use the projection "map" $Z \subset \mathbb{P}^n \times \mathbb{P}^{n-1}$ of (2.21) and the diagram:

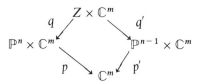

We may assume p' closed by induction, so it suffices to prove q' closed. And q' is seen to be a $\mathbb{P}^1$-bundle; in fact

$$q^{-1}((\mathbb{P}^{n-1} - H_i) \times \mathbb{C}^m) \cong \mathbb{P}^1 \times [(\mathbb{P}^{n-1} - H_i) \times \mathbb{C}^m].)$$

Secondly, if $S \subset \mathbb{P}^1 \times \mathbb{C}^m$ is defined by

$$\begin{aligned} &f_i(Z_1,\ldots,Z_n;X,Y) = 0, \quad 1 \leq i \leq t \\ &f_i \text{ homog. in } X, Y \text{ of degree } d \\ &Z_i \text{ coordinates on } \mathbb{C}^m, \end{aligned}$$

look at the resultant of the two polynomials *in X*:

$$R(\Sigma t_i f_i(Z;X,1), \Sigma s_i f_i(Z;X,1))$$

and expand it as a polynomial in the t_i and s_i:

$$R = \sum R_{\alpha\beta}(Z) \cdot t^\alpha \cdot s^\beta.$$

Then it is easy to see that the equations $R_{\alpha\beta}(Z) = 0$, all α, β define the projection $p_2(S)$.

(2.26) Corollary. *Let $X \subset \mathbb{P}^n$ and $Y \subset \mathbb{P}^m$ be varieties. Let $Z \subset X \times Y$ be a correspondence. Then for all Zariski closed sets $S \subset X$,*

$$Z[S] = \{y \in Y \mid \exists x \in S \text{ such that } (x,y) \in Z\}$$

is a closed subset of Y.

(2.27) Corollary. *Let X, Y and Z be three projective varieties. Let $A \subset X \times Y$ and $B \subset Y \times Z$ be two correspondences. Then*

$$B \circ A = \{(x,z) \in X \times Z \mid y \in Y \text{ such that } (x,y) \in A, (y,z) \in B\}$$

is a correspondence from X to Z.

   Proof. Let $p_{13} : X \times Y \times Z \longrightarrow X \times Z$ be the projection. Then by definition, $B \circ A = p_{13}(A \times Z \cap X \times B)$. Since p_{13} is closed by (2.23) and $(A \times Z) \cap (X \times B)$ is closed in $X \times Y \times Z$, the result follows.
   Applying (2.23) to the projection map of (2.21), we get

(2.28) Corollary. *Let $p : \mathbb{P}^n - \{P_n\} \longrightarrow \mathbb{P}^{n-1}$ be the projection map of (2.21). Let $X \subset \mathbb{P}^n$ be a subvariety such that $P_n \notin X$. Then the restriction*

$$\bar{p} : X \longrightarrow \mathbb{P}^{n-1}$$

is a closed map: in particular $X' = p(X)$ *is a projective variety. Moreover* $\bar{p}: X \longrightarrow X'$ *has finite fibres,* $\dim X = \dim X'$, *and in the ring inclusion*

$$\mathbb{C}[X_0,\ldots,X_n]/I(X) \supset \mathbb{C}[X_0,\ldots,X_{n-1}]/I(X'),$$

the homogeneous coordinate ring of X *is a finitely generated module over the homogeneous coordinate ring of* X'.

Proof. The first part is clear. Next, the fibres $p^{-1}(y)$ of the projection are lines l_y through P_n, minus P_n. But $l_y \subset X$ implies $P_n \in X$; therefore $l_y \cap X$ is a proper closed algebraic subset of l_y, hence is a finite set. To prove the last part, since $P_n \notin X$, the ideal $I(X)$ of X contains a homogeneous polynomial f of the form:

$$f = X_n^d + a_1(X_0,\ldots,X_{n-1})X_n^{d-1} + \ldots + a_d(X_0,\ldots,X_{n-1}).$$

This means that $\{1, X_n, X_n^2, \ldots, X_n^{d-1}\}$ generate the homogeneous coordinate ring of X, as a module over the subring generated by $X_0,\ldots,X_{n-1}$, namely the homogeneous coordinate ring of X'. It also shows that these two rings have the same transcendence degree, hence $\dim X = \dim X'$.   QED

We can compose a finite set of projections with each other, leading to the following general class of maps:

let $Y_i = \displaystyle\sum_{j=0}^{n} a_{ij}X_j$, $0 \leq i \leq r$,   be $(r+1)$ independent linear forms

let $L \subset \mathbb{P}^n$ be the $(n-r-1)$-dimensional linear space $V(Y_0,\ldots,Y_r)$.

Define

$$p_L : \mathbb{P}^n - L \longrightarrow \mathbb{P}^r$$

by

$$(b_0,\ldots,b_n) \longrightarrow (\Sigma a_{0j}b_j,\ldots,\Sigma a_{rj}b_j).$$

p_L is called the *projection with center* L.

(2.29) Corollary. (Noether Normalization Lemma). *Let* X *be an* r-*dimensional subvariety of* $\mathbb{P}^n$. *Then there is a linear subspace* L *of dimension* $n-r-1$ *such that* $L \cap X = \phi$. *For all such* L, *the projection* p_L *restricts to a finite-to-one surjective closed map:*

$$p_L : X \longrightarrow \mathbb{P}^r.$$

and the homogeneous coordinate ring $\mathbb{C}(X_0,\ldots,X_n]/I(X)$ *of* X *is a finitely generated module over* $\mathbb{C}[Y_0,\ldots,Y_r]$.

Proof. If $r = n$, then $X = \mathbb{P}^n$ and there is nothing to prove. Use induction on $n - r$. If $r < n$, choose $x \in \mathbb{P}^n - X$ and let $X' = p_x(X) \subset \mathbb{P}^{n-1}$. Then (2.29) is true for X'. Then if $M \subset \mathbb{P}^{n-1}$ is an $(n-r-2)$-dimensional linear subspace disjoint from X',

$$L = \{x\} \cup p_x^{-1}(M) \subset \mathbb{P}^n$$

is an $(n-r-1)$-dimensional linear subspace disjoint from X. But if L was given, choose any $x \in L$, and then $M = p_x(L - \{x\})$ is an $(n - r - 2)$-dimensional linear subspace of $\mathbb{P}^{n-1}$ disjoint from X' and $p_L = p_M \circ p_x$. Then the rest follows by induction.   QED

The reader should realize that projective space is essential for (2.23): the projection

$$p_2 : \mathbb{C}^n \times \mathbb{C}^m \longrightarrow \mathbb{C}^m$$

is *not* closed in the Zariski topology. For instance look at:

$$\mathbb{C}^3_{X,Y,Z} \supset V(XZ - Y)$$
$$\downarrow$$
$$\mathbb{C}^2_{X,Y}$$

The image of the surface $Y = XZ$ is the union of the Zariski-open set $\{(x,y) \mid x \neq 0\}$ and of the origin $(0,0)$:

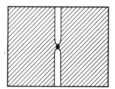

This is a typical constructible set:

(2.30) **Definition.** *A subset S of $\mathbb{C}^n$ or $\mathbb{P}^n$ is constructible if it is in the Boolean algebra generated by the closed algebraic sets; or equivalently if S is a disjoint union $T_1 \cup \ldots \cup T_k$, where T_i is locally closed, i.e., $T_i = T_i' - T_i''$, T_i' a closed algebraic set and $T_i'' \subset T_i'$ a smaller closed algebraic set.*

What is true for affine projections is:

(2.31) **Proposition.** *Let $S \subset \mathbb{C}^n \times \mathbb{C}^m$ be a constructible set. Then $p_2(S) \subset \mathbb{C}^m$ is a constructible set. In particular, if S is a subvariety of $\mathbb{C}^{n+m}$ and $\overline{p_2(S)}$ is the Zariski-closure of the image, then $p_2(S)$ contains a Zariski-open set in $\overline{p_2(S)}$.*

Proof. By taking compositions, we are reduced to the case of the projection

$$p_2 : \mathbb{C} \times \mathbb{C}^m \longrightarrow \mathbb{C}^m.$$

By induction on the dimension of $\overline{p_2(S)}$, we see easily that it is enough to prove the special case:

    if $S \subset \mathbb{C}^{m+1}$ is a subvariety and $S_0 \subset S$ is Zariski-open,

    then $p_2(S_0)$ contains a Zariski-open subset of $\overline{p_2(S)}$.

Let $T = \overline{p_2(S)}$. Then T is a variety and the affine rings of S and T satisfy:

$$\mathbb{C}[X_1, \ldots, X_{m+1}]/I(S) \longleftarrow \mathbb{C}[X_2, \ldots, X_{m+1}]/I(T)$$
$$\|\qquad\qquad\qquad\qquad\qquad\qquad\qquad \|$$
$$R_S \qquad\qquad\qquad\qquad\qquad\qquad\qquad R_T$$

(because $p_2(S)$ is Zariski dense in T hence no $f \in R_T, f \not\equiv 0$ can vanish on $p_2(S)$).
Thus $R_S \cong R_T[X_1]/\mathfrak{A}$ for some ideal $\mathfrak{A}$. We distinguish two cases: $\mathfrak{A} = (0)$ hence
$S = \mathbb{C} \times T; \mathfrak{A} \neq (0)$, hence dim $S = $ dim T. In the first case, let $(\alpha, t) \in S_0$.
Then $S_0 \cap \{\alpha\} \times T$ contains a Zariski-open subset $\{\alpha\} \times T_0$ of $\{\alpha\} \times T$, hence
$p_2(S_0) \supset T_0$. In the second case, let $\bar{S}$ be the Zariski-closure of S in $\mathbb{P}^1 \times \mathbb{C}^m$. Let
$S^* = [\bar{S} \cap (\{\infty\} \times \mathbb{C}^m)] \cup [S - S_0]$. Then S^* is a proper closed subset of $\bar{S}$, hence
dim(all comp. of S^*) < dim S. Clearly

$$p_2(S_0) \supset p_2(\bar{S}) - p_2(S^*)$$

and since $p_2(\bar{S})$ is closed by (2.23), $p_2(\bar{S}) = \overline{p_2(S)} = T$. But no component S^{**} of
S^* could map onto T because if it did, the affine ring R_T of T would inject into
some affine ring of S^{**} (with coordinates $X_1, \ldots, X_{m+1}$ or $\dfrac{1}{X_1}, X_2, \ldots, X_{m+1}$) and
then we would get dim $S^{**} \geq$ dim $T = $ dim S which is false. So $p_2(S^*)$ is a proper
closed subset of T and $p_2(S_0) \supset T - p_2(S^*)$.   QED

We want to apply (2.29) to show that non-empty Zariski-open subsets of a
variety are not only Zariski-dense but even classical dense in this variety. However,
we need a slight refinement of (2.29) first:

(2.32) **Proposition.** *Given* $a \in X \subset \mathbb{P}^n$, dim $X = r$, *then the projection* p_L *of* (2.29)
can be factored:

$$
\begin{array}{ccc}
\mathbb{P}^n - M & \xrightarrow{\;\;p_M\;\;} & \mathbb{P}^{r+1} \\
\cup & & \cup \\
 & & \mathbb{P}^{r+1} - \{x\} \xrightarrow{\;\;p_x\;\;} \mathbb{P}^r \\
 & & \cup \\
X & \xrightarrow{\;q = res_x(p)\;} & p_M(X) \nearrow
\end{array}
$$

(i.e., dim $M = n - r - 2$, $X \cap M = \phi$ and $x \notin p_M(X)$) where $q^{-1}q(a) = \{a\}$.

Proof. In $\mathbb{P}^n$, let L^* be the $(n - r)$-dimensional linear space which is the "join"
of L and $\{a\}$, i.e., the set of points x with homogeneous coordinates $y + \lambda \cdot \underline{a}$, y
homogeneous coordinates for a point of L, $\underline{a}$ homogeneous coordinates of a. Then
for all $x \in \mathbb{P}^n - L$

$$
\begin{array}{rl}
x \in L^* & \Longleftrightarrow p_L(x) = p_L(a) \\
 & \Longleftrightarrow x \in p_L^{-1}(p_L a).
\end{array}
$$

Let $p_L^{-1}(p_L(a)) \cap X = \{a, b_1, \ldots, b_k\}$. Let L_0^* be a hyperplane in L^*, i.e., a sublinear
space of dimension $n - r - 1$, such that $a \in L_0^*$, $b_i \notin L_0^*$ all i. Let $M = L_0^* \cap L$: this
is a linear space of dimension $n - r - 2$, whose join with a is just L_0^*. Use this M
to define p_M as in the above diagram. Then

$$
\begin{array}{rl}
x \in L_0^* \cap X & \Longleftrightarrow p_M(x) = p_M(a) \text{ and } x \in X \\
 & \Longleftrightarrow x \in p_M^{-1}(p_M(a)) \cap X,
\end{array}
$$

and by construction a is the only point in $L_0^* \cap X$.   QED

(2.33) **Theorem.** *Let* $X \subset \mathbb{P}^n$ *be an r-dimensional variety and let* X_0 *be a Zariski-
open set in* X. *Then the closure of* X_0 *in the classical topology is* X.

Proof.* Let $c \in X - X_0$. We will show that c is the limit in the classical topology of points of X_0. Choose M and x as in (2.32). Choose coordinates $X_0, \ldots, X_n$ on $\mathbb{P}^n$ such that

$$1) \quad M = V(X_0, \ldots, X_{r+1})$$

$$2) \quad c = (1, 0, \ldots, 0)$$

$$3) \quad M \cup p_M^{-1}(x) = V(X_0, \ldots, X_r).$$

Then $X_0, \ldots, X_{r+1}$ are coordinates on $\mathbb{P}^{r+1}$ and $X_0, \ldots, X_r$ are coordinates on $\mathbb{P}^r$. Now $p_x \circ p_M(X - X_0)$ is a closed algebraic subset of $\mathbb{P}^r$ of dimension at most $r - 1$. Choose a homogeneous polynomial $f(X_0, \ldots, X_r)$ such that $p_x \circ p_M(X - X_0) \subset V(f)$, or equivalently $X_0 \supset X_0' \overset{\text{def}}{=} \{x \in X \mid f(x) \neq 0\}$. We may as well replace X_0 by X_0' and show $c \in$ classical closure (X_0'). Next, $p_M(X)$ is an r-dimensional subvariety of $\mathbb{P}^{r+1}$ so it is a hypersurface $F(X_0, \ldots, X_{r+1}) = 0$.

Step I. Since $f \not\equiv 0$, $\exists (a_0, \ldots, a_r)$ such that $f(a) \neq 0$. Look at the combinations $(1, 0, \ldots, 0) + t(a_0, \ldots, a_r)$ of $\varepsilon_0 + ta$ for short. Since $f(\varepsilon_0 + t \cdot a)$ is a polynomial in t which is not identically zero, it has only a finite set of zeroes. Therefore there exists a sequence $t_i \in \mathbb{C}, t_i \to 0$ as $i \to \infty$ such that $f(\varepsilon_0 + t_i a) \neq 0$, all i. Thus $(\varepsilon_0 + t_i a) \in (\text{open set } f \neq 0)$, and $(\varepsilon_0 + t_i a) \to \varepsilon_0 = p_x p_M(c)$ as $i \to \infty$.

Step II. Write out F:

$$F(X_0, \ldots, X_{r+1}) = \alpha X_{r+1}^d + a_1(X_0, \ldots, X_r) X_{r+1}^{d-1} + \ldots + a_d(X_0, \ldots, X_r).$$

Since $x \in \mathbb{P}^{r+1}$ has coordinates $(0, , \ldots, 0, 1)$ and $F(x) \neq 0$, it follows that $\alpha \neq 0$. Moreover since $p_M(c) \in p_M(X)$, it follows that $0 = F(1, \ldots, 0, 0) = a_d(\varepsilon_0)$. We want to find points $b_i \in \mathbb{P}^{r+1}$ such that $F(b_i) = 0, p_x(b_i) = \varepsilon_0 + t_i a$, and $b_i \to (1, 0, \ldots, 0) = p_M(c)$ as $i \to \infty$. If we write

$$b_i = (1 + t_i a_0, t_i a_1, \ldots, t_i a_r, \beta_i),$$

then β_i is to be a root of

$$F_i(X) = \alpha X^d + a_1(\varepsilon_0 + t_i a) X^{d-1} + \ldots + a_d(\varepsilon_0 + t_i a).$$

To ensure that $b_i \to (1, 0, \ldots, 0)$, we want $\beta_i \to 0$. But the product of the roots of F_i is $a_d(\varepsilon_0 + t_i a)/\alpha$, and since $a_d(\varepsilon_0) = 0$, these numbers tend to 0. Therefore there is indeed a sequence β_i of roots that tends to 0.

Step III. Lift b_i to any $c_i \in X$. Now since $\mathbb{P}^n$ is compact in the classical topology, so is X and some subsequence c_{i_k} must converge. Let c_∞ be its limit. Then

$$p_M(c_\infty) = \lim_{i \to \infty} p_M(c_i) = \lim_{i \to \infty} b_i = p_M(c).$$

Since $p_M^{-1} p_M(c) = \{c\}$, it follows that $c_\infty = c$. Since $f(c_i) \neq 0$. this shows that $c \in$ classical closure (X_0').   QED

*Following suggestions of G. Stolzenberg.

Chapter 3. Structure of Correspondences

§3A. Local Properties—Smooth Maps, Fundamental Openness Principle, Zariski's Main Theorem

The goal of this section is to analyze the fibres of a projection

$$Z \subset \mathbb{P}^n \times \mathbb{P}^m$$
$$\downarrow \quad \downarrow {\scriptstyle p_2}$$
$$X = p_2(Z) \subset \mathbb{P}^m$$

and hence to understand some of the qualitative properties of correspondences. The first part of our analysis can be carried out locally on Z and X, i.e., it concerns a projection

$$Z \subset \mathbb{C}^{n+m}$$
$$\downarrow \quad \downarrow {\scriptstyle p_2}$$
$$X = \overline{p_2(Z)} \subset \mathbb{C}^m$$

But, in fact, it seems more natural when we are studying affine varieties to look at a general *regular map* of affines:

$$X \xrightarrow{\quad \varphi \quad} Y$$
$$\cap \qquad\qquad \cap$$
$$\mathbb{C}^n \qquad\qquad \mathbb{C}^m$$

(coordinates $X_1,\dots,X_n$) (coordinates $Y_1,\dots,Y_m$)

By definition, a map φ is called regular if φ is defined by polynomials: $Y_i = \varphi_i(X_1,\dots,X_n)$, $1 \leq i \leq m$, $\varphi_i \in \mathbb{C}[X_1,\dots,X_n]$. We say that φ is *dominating* if $\varphi(X)$ is Zariski-dense in Y, i.e., $Y = \overline{\varphi(X)}$. We are not really introducing any greater generality into the discussion though, because any regular map φ can be factored:

$$\mathbb{C}^{n+m}$$
$$\cup$$
$$Z = \text{locus of points } \{x, \varphi(x)\}, x \in X$$

(3.1)   bijection $\diagup$ $\diagdown$ projection

$$\mathbb{C}^n \supset X \xrightarrow{\qquad\qquad} Y \subset \mathbb{C}^m$$
$$\varphi$$

where Z is, in fact, a variety, being defined by the equations $f_i(X_1,\dots,X_n) = 0$, (all $f_i \in$ ideal of X) and $Y_i - \varphi_i(X_1,\dots,X_n) = 0$, $1 \leq i \leq m$. For the whole of §3A, we will study a regular map $\varphi : X \to Y$ of affine varieties.

First of all, we want to generalize some of the results of Ch. 1 on affine varieties to "relative" versions involving regular maps. Let $\varphi : X \to Y$ be a regular map.

(3.2) **Definition.** *If $x \in X$ and $y = \varphi(x)$, then $d\varphi : T_{x,X} \to T_{y,Y}$ is the linear map defined by:*

$$
\left.
\begin{array}{l}
\textit{Given } (\xi_1, \ldots, \xi_n) \in T_{x,X}, \textit{ i.e. } \xi_i \\
\textit{satisfying the linear equations} \\[1ex]
\displaystyle \sum_{i=1}^{n} \frac{\partial f}{\partial X_i}(x) \cdot \xi_i = 0, \\[2ex]
\textit{all } f \in I(X)
\end{array}
\right\}
\quad \textit{let } d\varphi(\xi) = \left(\sum \frac{\partial \varphi_1}{\partial X_i}(x)\xi_i, \ldots, \sum \frac{\partial \varphi_m}{\partial X_i}(x)\xi_i \right).
$$

$$
\left[\textrm{Check that } \sum_{j=1}^{m} \frac{\partial g}{\partial Y_j}(y) \cdot \left(\sum_{i=1}^{n} \frac{\partial \varphi_j}{\partial X_i}(a) \cdot \xi_i \right) = 0, \quad \textrm{all } g \in I(Y), \textrm{ so that } d\varphi(\xi) \in T_{y,Y} \, ! \right]
$$

Note that $g \to \varphi^* g = g \circ \varphi, g \in \mathbb{C}(Y_1, \ldots, Y_m]$, defines a homomorphism:

$$
R_X \xleftarrow{\quad \varphi^* \quad} R_Y
$$

of the affine coordinate rings. If φ is dominating, then φ^* is injective, hence it extends to an injection of function fields:

$$
\mathbb{C}(X) \longleftarrow \!\!\!\!\!\!\!\!\supset \mathbb{C}(Y).
$$

In particular if $r = \dim X, s = \dim Y$, then $r \geq s$.

(3.3) **Definition.** *Let $\varphi : X^r \to Y^s$ be a dominating regular map. If $x \in X$ and $y = \varphi(x) \in Y$ is a smooth point then we say that φ is smooth at x if x is a smooth point of X and $d\varphi$ maps $T_{x,X}$ onto $T_{y,Y}$. Since by assumption $\dim T_{y,Y} = s$ and $\dim T_{x,X} \geq r$, this is equivalent to saying that $\ker(d\varphi)$ is $r - s$-dimensional.*

Concerning this definition, we get the following Propositions by direct analogy with the results of Chapter 1 (which treat the case $Y =$ one point):

(3.4) **Proposition.** *If φ is smooth at x, then $\varphi^{-1}\varphi(x)$ has a unique component Z through x, $\dim Z = r - s$, and x is a smooth point of Z.*

(3.5) **Proposition.** *If φ is smooth at x, then there are open neighborhoods $U \subset X$, $V \subset Y$ of x and $\varphi(x)$ in the classical topology and analytic coordinates α and β in U and V:*

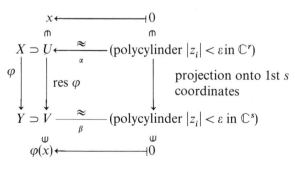

where $\varphi(U) \subset V$ and φ goes over in the coordinates to projection.

(3.6) **Proposition.** *The set of smooth points of φ is a non-empty Zariski-open subset of X.*

Sketch of proofs. (3.4) is easy—if X^r is in $\mathbb{C}^n$, then near x, X is defined by exactly $n - r$ equations f_i with independent differentials. Since $\varphi(x)$ is smooth on Y^s, choose s equations $g_1, \ldots, g_s \in \mathbb{C}[Y_1, \ldots, Y_m]$ such that $g_i(\varphi(x)) = 0$ and the g_i's have independent differentials in $T_{\phi(x),Y}$. Then the one-point subvariety $\{\varphi(x)\} \subset Y$ is one of the components of $\{y \in Y \,|\, g_1(y) = \ldots = g_s(y) = 0\}$ by (1.16). Therefore $\varphi^{-1}\varphi(x)$ is defined in a Zariski-open neighborhood of x by $n - r + s$ equations $f_1 = \ldots = f_{n-r} = \varphi^*g_1 = \ldots = \varphi^*g_s = 0$, with independent differentials. By (1.16) this proves (3.4). As for (3.5), this is a typical application of the convergent power series implicit function theorem. Take $g_1, \ldots, g_s$ as above: then $g_1, \ldots, g_s$ define analytic coordinates in a neighborhood of $\varphi(x) \in Y$. Let $h_1, \ldots, h_{r-s} \in \mathbb{C}[X_1, \ldots, X_n]$ vanish at x and have differentials which are independent linear forms on the subspace $\ker(d\varphi) \subset T_{x,X}$. Then $h_1, \ldots, h_{r-s}, \varphi^*g_1, \ldots, \varphi^*g_s$ have independent differentials at x, hence define analytic coordinates in a neighborhood of x. To prove (3.6), let $X = V(f_1, \ldots, f_k)$ and consider the matrix of polynomials:

$$M = \begin{pmatrix} \partial f_1/\partial X_1 \ldots \partial f_k/\partial X_1 \; \partial \varphi_1/\partial X_1 \ldots \partial \varphi_m/\partial X_n \\ \cdot \qquad\qquad\qquad \cdot \qquad\qquad \cdot \\ \partial f_1/\partial X_n \ldots \partial f_k/\partial X_n \; \partial \varphi_1/\partial X_n \ldots \partial \varphi_m/\partial X_n \end{pmatrix}$$

As x ranges over $X - \varphi^{-1}(\text{Sing } Y)$, rank $M(x)$ is always at most $n - r + s$, and rank $M(x) = n - r - s$ if and only if φ is smooth at x. On the other hand, if we consider M as a matrix of elements of $\mathbb{C}(X)$, it defines a linear transformation

$$\mathbb{C}(X)^n \xrightarrow{\;\;\widetilde{M}\;\;} \mathbb{C}(X)^{k+m}$$

whose kernel is in one-to-one correspondence with the vector space of derivations $D : \mathbb{C}(X) \to \mathbb{C}(X)$ such that $D \equiv 0$ on $\varphi^*\mathbb{C}(Y)$ via:

$$\begin{cases} D \text{ defined by elements } \delta_i = D(X_i) \\ \delta_1, \ldots, \delta_n \text{ define a } D \text{ if and only if} \\ \displaystyle\sum_{j=1}^n \frac{\partial f_{i_1}}{\partial X_j} \delta_j = \sum_{j=1}^n \frac{\partial \varphi_{i_2}}{\partial X_j} \delta_j = 0, \quad \text{all } i_1, i_2 \end{cases}$$

Since $\text{tr.d.}_{\phi^*\mathbb{C}(Y)}\mathbb{C}(X) = r - s$, the vector space of these derivations **has dimension** $r - s$ and so rank $\widetilde{M} = n - r + s$. As in §1A this implies that rank $M(x) = n - r - s$ for almost all x.   QED

(3.6) can be significantly strengthened in one way:

(3.7) **Sard's lemma for varieties.** *$\exists$ a Zariski open set $Y_0 \subset Y$ such that φ is smooth at all points $x \in \varphi^{-1}(Y_0) - \text{Sing } X$.*

Proof. Let X_0 be the sets of points where φ is smooth and let $W = X - X_0 - \text{Sing } X - \varphi^{-1}(\text{Sing } Y)$: a locally closed subset of X. We want to show that $\overline{\varphi(W)} \subsetneqq Y$. Suppose in fact that $\overline{\varphi(W)} = Y$. Then $\overline{W}$ has a component $\widehat{W}_1$ such that $\psi = \text{res } \varphi : \overline{W}_1 \to Y$ is a dominating regular map. By (3.6), there are points in the

Zariski open set $W \cap \bar{W}_1$ of $\bar{W}_1$ such that ψ is smooth at x. But then res $d\varphi = d\psi$ maps $T_{x,\bar{W}_1}$ onto $T_{\varphi(x),Y}$, hence $d\varphi$ also maps the larger vector space $T_{x,X}$ onto $T_{\phi(x),Y}$. Since $x \in W$, x and $\varphi(x)$ are smooth, so this shows that φ is smooth, hence $x \notin W$: contradiction.

(3.8) **Corollary.** $\exists$ *a Zariski open set* $Y_0 \subset Y$ *such that* $\varphi^{-1}(Y_0)$ *can be stratified into pieces smooth over* Y_0, *i.e.,*

$$\varphi^{-1}(Y_0) = X_0 \cup X_1 \cup \ldots \cup X_k, X_i \text{ disjoint locally closed subsets of } X$$

such that res $\varphi: X_i \to Y_0$ *is smooth for all i, and* $\dim X_0 = \dim X$, *but* $\dim X_i < \dim X$ $(i > 0)$.

Proof. Stratify X as in §1A and apply (3.7) to each piece.

We now proceed to a much deeper and more important result. This is the fundamental openness result in complex analytic geometry as well as algebraic geometry:

(3.9) **Definition.** *Let* X *be an affine variety and* $x \in X$. *Then* X *is topologically unibranch at* x *if for all closed algebraic subsets* $Y \subsetneqq X$, x *has a fundamental system of neighborhoods* $\{U_n\}$ *in the classical topology on* X *such that* $U_n - U_n \cap Y$ *is connected.*

Note that smooth points are topologically unibranch: in fact, if x is smooth on X, x has analytic coordinates in a neighborhood U:

$$
\begin{array}{ccc}
x & \longleftarrow & 0 \\
\rotatebox{90}{$\in$} & & \rotatebox{90}{$\in$} \\
X \supset U & \underset{\alpha}{\overset{\approx}{\longleftarrow}} & \left(\begin{matrix} \text{polycylinder } V_\varepsilon \text{ defined by} \\ |z_i| < \varepsilon \text{ in } \mathbb{C}^r. \end{matrix} \right)
\end{array}
$$

Every $Y \subsetneqq X$ is contained in the set of zeroes of some polynomial p which doesn't vanish identically on X; hence $\alpha^{-1}(Y)$ is contained in the set of zeroes of some convergent power series $q(z_1, \ldots, z_n)$ which doesn't vanish identically. And V_ε-(zeroes of q) is certainly connected. [If $x, y \in V_\varepsilon$-(zeroes of q), look at the line $tx + (1-t)y$ joining them (vector notation). Then $tx + (1-t)y \in V_\varepsilon$ if $|t - \frac{1}{2}| \leq \frac{1}{2}$ and $q(tx - (1-t)y)$ has only a finite set of zeroes on the disc $|t - \frac{1}{2}| \leq \frac{1}{2}$. Thus x and y can be joined by a path avoiding the zeroes of q.] More generally, this argument shows that to check X topologically unibranch at x, it suffices to take $Y = \text{Sing } X$ in the definition. In fact, if $U - U \cap \text{Sing } X$ is a connected r-dimensional analytic manifold, then for any $Y \subsetneqq X$, the above argument shows that $U - U \cap (\text{Sing } X \cup Y)$ is still connected, and since it is dense in $U - U \cap Y$ by (2.33), $U - U \cap Y$ is connected.

(3.10) **Fundamental Openness Principle.** *Let* $\varphi: X^r \to Y^r$ *be a regular dominating map of affine varieties of the same dimension. Let* $x \in X$ *be such that*
  a) $\varphi(x)$ *is topologically unibranch on* Y
  b) $\{x\}$ *is a component of* $\varphi^{-1}\varphi(x)$.
Then φ *is open at* x *in the classical topology, i.e., for all neighborhoods* $U \subset X$ *of* x, $\varphi(U)$ *contains a neighborhood of* $\varphi(x)$ *in* Y.

Proof. Step I: There are open neighborhoods $U \subset X, V \subset Y$ of x and $y = \varphi(x)$ such that $\varphi(U) \subset V$ and res $\varphi: U \to V$ is proper, ("proper" means inverse images of compact sets are compact) in the classical topology. This follows from a quite general fact:

(3.11) **Lemma.** *Let $f: X \to Y$ be a continuous map of locally compact topological spaces. Let $y \in Y$ and assume $f^{-1}(y)$ is compact. Then there are open neighborhoods $U \subset X, V \subset Y$ of $f^{-1}(y)$ and y such that $f(U) \subset V$ and res $f: U \to V$ is proper.*

Proof. Let X_0 be an open neighborhood of $f^{-1}(y)$ such that $\bar{X}_0$ is compact. For all open neighborhoods $Y_\alpha \subset Y$ of y, suppose res $f: X_0 \cap f^{-1}(Y_\alpha) \to Y_\alpha$ is not proper. But res $f: \bar{X}_0 \cap f^{-1}(Y_\alpha) \to Y_\alpha$ is proper, so $\bar{X}_0 \cap f^{-1}(Y_\alpha) \supsetneq X_0 \cap f^{-1}(Y_\alpha)$. Let $x_\alpha \in \bar{X}_0 - X_0$ be such that $f(x_\alpha) \in Y_\alpha$. As the Y_α shrink, $f(x_\alpha) \to y$. Since X_0 is compact, the $\{x_\alpha\}$'s have a point x_∞ of accumulation. Then on the one hand $x_\infty \in \bar{X}_0 - X_0$, while on the other hand $f(x_\infty) = y$, hence $x_\infty \in f^{-1}(y) \subset X_0$.

Step II: Let $J = $ locus of points $x \in X$ such that either x or $\varphi(x)$ is singular or φ is not smooth. J is a closed algebraic set all of whose components have lower dimension than X. Let $B = \overline{\varphi(J)}$. B is a closed algebraic set in Y all of whose components have lower dimension than X, hence lower than that of Y too. Thus $B \subsetneq Y$.

Step III. By the top. unibranch hypothesis, choose a neighborhood $V' \subset V$ of $\varphi(x)$ such that $V' - B \cap V'$ is connected. Let $U' = U \cap \varphi^{-1}(V')$. Look at the map

$$\psi = \text{res } \varphi : U' - \varphi^{-1}(B) \to V' - B.$$

It is smooth and proper. By (3.5), it is a local homeomorphism. Hence by the theory of covering spaces

$$y \longrightarrow \# \psi^{-1}(y)$$

is a locally constant function on $V' - B$. Since $V' - B$ is connected it follows that $y \to \# \psi^{-1}(y)$ is a constant. Now $U' - \varphi^{-1}B \neq \phi$ because $x \in$ (closure of $X - \varphi^{-1}B$ in the classical topology): (2.33). Thus $\# \psi^{-1}(y) > 0$ for some $y \in V' - B$. Thus $\# \psi^{-1}(y) > 0$, all $y \in V' - B$, and $V' - B \subset \psi(U')$. But $\varphi: U' \to V'$ is proper so $\varphi(U')$ is closed in V'. Again by (2.33) the classical closure of $V' - B$ is V'. So $\varphi(U') = V'$.
                                                                                              QED

This theorem points out the big difference between real and complex analytic geometry. Over the reals, the map $\varphi(x) = x^2$ from $\mathbb{R}$ to $\mathbb{R}$ is not open at $x = 0$, for instance, but over the complexes, $\varphi(x) = x^2$ from $\mathbb{C}$ to $\mathbb{C}$ is clearly open. An example showing that the hypothesis $\varphi(x)$ unibranch is essential will be given below (c.f. end of §3B). Moreover the proof of the theorem shows that we can make a very important definition:

(3.12) **Definition.** *Let $\varphi: X^r \to Y^r$ be a regular dominating map of affine varieties of the same dimension. Let $x \in X$ satisfy:*
   a) *$\varphi(x)$ is topologically unibranch on Y,*
   b) *$\{x\}$ is a component of $\varphi^{-1}\varphi x$.*
Choose open neighborhoods $x \in U \subset X$ and $\varphi x \in V \subset Y$ and a closed algebraic subset $B \subset Y$ such that

   c) $V - V \cap B$ is conntected
   d) $\varphi(U) \subset V$, res $\varphi : U \to V$ is proper and $\varphi^{-1}\varphi x \cap U = \{x\}$,
   e) res $\varphi : U - \varphi^{-1}B \to V - B$ is smooth.
Then let

$$\text{mult}_x(\varphi) = \#(\varphi^{-1}y \cap U), \text{ any } y \in V - B.$$

(It is immediate that this number is independent of the choice of U, V and B.)
In other words, we can measure quite precisely how many "branches" of the
covering X over Y come together at x. The theorem has moreover many important
qualitative applications:

(3.13) Theorem. *Let* $\varphi : X^r \to Y^s$ *be a dominating regular map of affine varieties.
For all* $y \in Y$, *all components of* $\varphi^{-1}(y)$ *have dimension at least* $r - s$.

Proof. Let W be a component of $\varphi^{-1}(y)$ of dimension $< r - s$. Let $x \in W, x \notin$
other components. Choose s polynomial functions $g_1, \dots, g_s$ on Y, zero at y, such
that $\{y\} = $ component of $V(g_1, \dots, g_s) \cap Y$. [This is possible: start with any $g_1 \not\equiv 0$
on Y, then choose a $g_2 \not\equiv 0$ on any component of $Y \cap V(g_1)$, etc.; apply (1.13).]
Similarly choose $r - s - 1$ functions f_i on X, zero at x, such that $\{x\} = $ component
of $V(f_1, \dots, f_{r-s+1}) \cap W$. Altogether the f's and $\varphi^* g$'s define

$$\psi : X \longrightarrow \mathbb{C}^{r-1}$$
$$z \longrightarrow (f_1(z), \dots, f_{r-s+1}(z), g_1(\varphi z), \dots, g_s(\varphi z))$$

such that $\{x\}$ is a component of $\psi^{-1}(0)$. This in itself would be a contradiction
to the theorem. Now choose the smallest k such that there exists a regular map

$$\psi : X \longrightarrow \mathbb{C}^k$$

for which $\{x\}$ is a component of $\psi^{-1}(0)$. Let U be a Zariski open set containing x
such that $\psi^{-1}(0) \cap U = \{x\}$. In this case, note that ψ is dominating. In fact, if
$Z = \overline{\psi(X)}$ had dimension $l < k$, choose l polynomial functions $h_1, \dots, h_l$ on Z such
that $(0) = $ component of $Z \cap V(h_1, \dots, h_l)$ and replace ψ by $H \circ \psi$, $H : \mathbb{C}^k \to \mathbb{C}^l$ being
defined by $(h_1, \dots, h_l)$. Thus ψ is dominating and if $Y_1, \dots, Y_k$ are coordinates in
$\mathbb{C}^k$, $\psi^* Y_1, \dots, \psi^* Y_k$ are independent transcendentals in $\mathbb{C}(X)$. Choose $f_{k+1}, \dots, f_r$
polynomial functions on X zero at x such that $\{\psi^* Y_i, f_j\}$ are a transcendence base
of $\mathbb{C}(X)$. Enlarge ψ to

$$\tilde{\psi} : X \longrightarrow \mathbb{C}^r$$
$$\tilde{\psi}(z) = (\psi(z), f_{k+1}(z), \dots, f_r(z)).$$

We can now apply Theorem (3.10) to $\tilde{\psi}$! It follows that $\tilde{\psi}$ is open at x. But

$$0 = \tilde{\psi}(x) \in \tilde{\psi}(U)$$

$$(0, a_{k+1}, \dots, a_n) \notin \tilde{\psi}(U), \text{ any } (a_{k+1}, \dots, a_n) \neq (0, \dots, 0)$$

since if $y \in U$:

$$\tilde{\psi}(y) = (0, a_{k+1}, \dots, a_n) \Rightarrow \psi(y) = 0$$
$$\Rightarrow y \in U \cap \psi^{-1}(0) = \{x\}$$
$$\Rightarrow a_i = f_i(y) = 0.$$

So $\tilde{\psi}$ is not, in fact, open!   QED

(3.14) Corollary. *If X is an r-dimensional variety and $f_1,\ldots,f_k$ are polynomial functions on X, then every component of $X \cap V(f_1,\ldots,f_k)$ has dimension $\geq r - k$.*

(3.15) Corollary. *If $\varphi : X^r \to Y^s$ is a dominating regular map, there is a Zariski open set $Y_0 \subset Y$ such that for all $y \in Y_0$, every component of $\varphi^{-1}(y)$ has dimension $r - s$.*

Proof. (3.8) shows that for small enough Y_0, every component of $\varphi^{-1}(y)$ is a union of smooth varieties of dimension $\leq r - s$, hence its dimension is $\leq r - s$; but (3.13) shows that their dimension is always $\geq r - s$.

(3.16) Corollary. *If $\varphi : X^r \to Y^s$ is a regular map, then*

$$x \longmapsto \max(\dim W) \quad \begin{bmatrix} W \text{ a component} \\ \text{of } \varphi^{-1}\varphi(x) \end{bmatrix}$$

is an upper semi-continuous function in the Zariski topology, i.e., $\forall\, k$,

$$S_k(\varphi) = \{x \in X \mid \exists \text{ a component of } \varphi^{-1}\varphi(x) \text{ of dimension} \geq k\}$$

is Zariski-closed.

Proof. Prove this by induction on $\dim Y$. If $\overline{\varphi(X)} \subsetneqq Y$, replace Y by $\overline{\varphi(X)}$ and we are through. If $\overline{\varphi(X)} = Y$, then choose Y_0 as in (3.15) and we find:

$$S_k(\varphi) = X \text{ if } k \leq r - s$$
$$S_k(\varphi) \subset \varphi^{-1}(Y - Y_0) \text{ if } k > r - s.$$

Breaking up $Y - Y_0$ into components $Y^{(1)}, \ldots, Y^{(k)}$ and breaking up $\varphi^{-1}(Y^{(i)})$ into components $X^{(i,1)}, \ldots, X^{(i,k_i)}$, and letting $\varphi_{ij} = \mathrm{res}\ \varphi : X^{(i,j)} \to Y^{(i)}$, it follows that

$$S_k(\varphi) = \bigcup_{i,j} S_k(\varphi_{ij}) \text{ if } k > r - s$$

which is closed by induction.   QED

A simple example where $\max(\dim W)$ is not continuous is given by:

$$\varphi : \mathbb{C}^2 \to \mathbb{C}^2$$
$$\varphi(a,b) = (a, ab)$$

In this case, $\varphi^{-1}\varphi(a,b) = \{(a,b)\}$ if $a \neq 0$, while $\varphi^{-1}\varphi(0,b) = \{$the line of all points $(0,c)\}$.

Next we want to look at the cardinality of the fibres of a map between two varieties of the same dimension:

(3.17) Proposition. *Let $\varphi : X^r \to Y^r$ be a dominating regular map. Then there is a Zariski-open $Y_0 \subset Y$ such that*

$$\# \varphi^{-1}(y) = [\mathbb{C}(X) : \mathbb{C}(Y)], \quad \text{all } y \in Y_0.$$

This number is called the degree of φ or $\deg(\varphi)$.

Proof. Say $X \subset \mathbb{C}^n$. Then $\mathbb{C}(X)$ is generated over $\mathbb{C}(Y)$ by the coordinate functions $X_1, \ldots, X_n$. By the theorem of the primitive element, some linear combination $l(X) = \sum \alpha_i X_i$ already generates $\mathbb{C}(X)$ over $\mathbb{C}(Y)$. Factor the map φ:

$$X \xrightarrow{\ \psi\ } Y \times \mathbb{C} \xrightarrow{\ p_1\ } Y$$
$$x \longrightarrow (\varphi(x), l(x))$$

and let $Z = \overline{\psi(X)}$. Then $\mathbb{C}(Z) \xrightarrow[\psi^*]{\approx} \mathbb{C}(X)$, and $[\mathbb{C}(X):\mathbb{C}(Y)] = [\mathbb{C}(Z):\mathbb{C}(Y)]$. For the mapping $\psi : X \to Z$, note that there are polynomial functions $f_0, \ldots f_n$ on Z such that $X_i = \psi^* f_i / \psi^* f_0$ in $\mathbb{C}(X)$, i.e., $\psi^* f_0 \cdot X_i - \psi^* f_i$ vanishes on X. It follows that if $z \in Z$ and $f_0(z) \neq 0$, then $\psi^{-1}(z)$ consists in at most one point, i.e., the point $\left(\dfrac{f_1(z)}{f_0(z)}, \ldots, \dfrac{f_n(z)}{f_0(z)} \right) \in \mathbb{C}^n$. Since $\psi^{-1}(z) \neq \phi$ for z in a suitable Zariski open, $\# \psi^{-1}(z) = 1$ for z in a suitable Zariski open. Now look at $p_1 : Z \twoheadrightarrow Y$. Let $n = [\mathbb{C}(Z):\mathbb{C}(Y)]$ and let $\sum\limits_{i=0}^{n} a_i T^i = 0$ be the irreducible equation satisfied in $\mathbb{C}(Z)$ by the last coordinate on Z (i.e., the coordinate in the 2nd factor of $Y \times \mathbb{C}$). We may assume that the a_i are polynomial functions on Y. Consider all solutions of this equation:

$$Z^* = \left\{ (y, t) \mid y \in Y, \sum a_i(y) t^i = 0 \right\} \subset Y \times \mathbb{C}.$$

Then I claim:

(3.18) $Z^* = Z \cup Z'$, where $\overline{p_1(Z')} \subsetneqq Y$.

In fact, if Z is defined by equations $f_i(Y, T) = 0$, then in $\mathbb{C}(Y)[T], f_i$ is a multiple of $\sum a_i T^i$, i.e.

$$b_i(Y) f_i(Y, T) \equiv c_i(Y, T) \cdot \sum_{j=0}^{n} a_j(Y) T^j + \text{something zero on } Y.$$

Therefore if $\prod\limits_i b_i(y) \neq 0$, every solution of the equation $\sum a_i(y) t^i = 0$, $y \in Y$, is a point of Z. This proves (3.18). It follows that for y in a suitable Zariski open set of Y,

$$\# (p_1^{-1}(y) \cap Z) \equiv [\text{number of solutions of } \sum a_i(y) T^i = 0].$$

Finally the derivative $\sum\limits_{i=1}^{n} i a_i T^{i-1}$ is a polynomial of lower degree in T, hence is not zero in $\mathbb{C}(Z)$. Therefore

$$\dim \left[Z \cap V \left(\sum_{1}^{n} i a_i T^{i-1} \right) \right] < \dim Y,$$

hence for almost all $y \in Y$, $\sum\limits_{i=0}^{n} a_i(y) T^i$ has all distinct roots. In this case, it has n roots and $\# (p_1^{-1}(y) \cap Z) = [\mathbb{C}(Z):\mathbb{C}(Y)]$ is proven.   QED

The same kind of ideas give us the following projective result:

(3.19) **Corollary.** *Let $X^r \subset \mathbb{P}^n$ be an r-dimensional variety. Then there exists a linear $L^{n-r-2} \subset \mathbb{P}^n$, $L \cap X = \phi$, such that the projection*

$$\text{res } p_L : X \longrightarrow p_L(X) \subset \mathbb{P}^{r+1}$$
$$\quad\quad\quad\quad\quad\quad\quad\quad \wr\|$$
$$\quad\quad\quad\quad\quad\quad V(F) \text{ hypersurface}$$

is birational.

Proof. Let M^{n-r-1} be a linear space disjoint from X defining a projection $p_M : X \longrightarrow \mathbb{P}^r$. p_M is surjective, hence it induces an inclusion of the function fields:

$$\mathbb{C}(X) \longleftarrow\!\!\!\supset \mathbb{C}(\mathbb{P}^r)$$

and since both have transcendence degree r, $\mathbb{C}(X)$ is a finite algebraic extension of $\mathbb{C}(\mathbb{P}^r)$. If $X_0, \dots, X_n$ are homogeneous coordinates on $\mathbb{P}^n$ such that $M = V(X_0, \dots, X_r)$, hence $(X_0, \dots, X_r)$ are coordinates on $\mathbb{P}^r$, then $\mathbb{C}(X)$ is generated over $\mathbb{C}(\mathbb{P}^r)$ by the images of the functions $\dfrac{X^{r+1}}{X_0}, \dots, \dfrac{X^n}{X_0}$. By the theorem of the primitive element, it is generated by a suitable linear combination $\displaystyle\sum_{i=r+1}^{n} \alpha_i\left(\dfrac{X_i}{X_0}\right)$. Let $L = M \cap V\left(\displaystyle\sum_{i=r+1}^{n} \alpha_i X_i\right)$. Then p_L lies in a diagram:

$$
\begin{array}{ccccc}
& & p_M & & \\
\mathbb{P}^n - M & \xrightarrow{\ p_L\ } & \mathbb{P}^{r+1} - \{x\} & \xrightarrow{\ p_x\ } & \mathbb{P}^r \\
\cup & & \cup & & \\
X & \longrightarrow & V(F) & &
\end{array}
$$

$$\left[\begin{array}{c} x = \text{image by } p_L \text{ of the} \\ \text{center } M \text{ of } p_M \end{array} \right]$$

where $V(F)$ is the hypersurface which is the image of X. We get inclusions of functions fields:

$$\mathbb{C}(X) \xleftarrow[\alpha]{\ \ \ \supset\ \ \ } \mathbb{C}(V(F)) \longleftarrow\!\!\!\supset \mathbb{C}(\mathbb{P}^r)$$
$$\quad\quad \| \quad\quad\quad\quad\quad\quad\quad \|$$
$$\mathbb{C}(\mathbb{P}^r)\left(\dfrac{X_{r+1}}{X_0}, \dots, \dfrac{X_n}{X_0}\right) \quad\quad \mathbb{C}(\mathbb{P}^r)\left(\dfrac{\sum \alpha_i x_i}{X_0}\right)$$

By assumption, the inclusion α is a surjection. Therefore by (3.17), res $p_L : X \to V(F)$ is almost everywhere one-to-one, hence is birational. QED

Concerning regular maps $\varphi : X^r \to Y^r$ which are birational, i.e., equivalently $\varphi^* : \mathbb{C}(Y) \to \mathbb{C}(X)$ is an isomorphism or φ is injective on a Zariski-open subset, we get the following very important result:

(3.20) Zariski's Main Theorem (smooth case). *Let $\varphi : X^r \to Y^r$ be a birational regular map. Let $X_1, \dots, X_n$, resp. $Y_1, \dots, Y_m$ be affine coordinates on X, resp. Y. Let $x \in X$ and assume $y = \varphi(x)$ is smooth on Y. Then either*

a) φ^{-1} is a regular correspondence at y; more precisely, $\exists$ a polynomial $d(Y_1, \dots, Y_m)$ such that $d(y) \neq 0$ and an inverse

$$\psi : Y' = \{y \in Y \,|\, d(y) \neq 0\} \xrightarrow{\hspace{2cm}} X$$

to φ defined by

$$X_i = \frac{a_i(Y_1, \ldots, Y_m)}{d(Y_1, \ldots, Y_m)},$$

or

b) *$\exists$ a subvariety $E \subset X$ through x of dimension $r - 1$ such that $\dim \overline{\varphi(E)} \leq r - 2$. Such an E is called an exceptional divisor. In particular, $\varphi^{-1}(y)$ has a positive-dimensional component through x.*

Proof. Since φ^* induces an isomorphism $\mathbb{C}(Y) \xrightarrow{\;\sim\;} \mathbb{C}(X)$, each function X_i on X equals $\dfrac{a_i(Y)}{b_i(Y)} \circ \varphi$ for some polynomials a_i, b_i ($b_i \not\equiv 0$ on Y). Using the fact that $\mathcal{O}_{y,Y}$ is a UFD, write:

$$X_i = \frac{a_i(Y)}{b_i(Y)} \circ \varphi, \text{ with } a_i, b_i \text{ relatively prime in } \mathcal{O}_{y,Y}.$$

There are 2 possibilities: either $b_i(y) \neq 0$, all i, or $b_i(y) = 0$ for some i. In the first case, let $d = \prod b_i$ and define ψ by

$$X_i = \frac{a_i(Y) \circ \prod_{j \neq i} b_j(Y)}{d(Y)}$$

In the second case, say $b_1(y) = 0$. Let $\beta(Y_1, \ldots, Y_m)$ be a polynomial that represents an irreducible factor of b_1 in $\mathcal{O}_{y,Y}$. Let E be a component of $X \cap V(\beta \circ \varphi)$ through x. By (3.14), $\dim E = r - 1$. But if $b_1 = b_1' \beta$, then on X, $a_1 \circ \varphi = X_1 \cdot (b_1' \circ \varphi) \cdot (\beta \circ \varphi)$, hence $a_1 \circ \varphi = 0$ on E. Therefore $a_1 = \beta = 0$ on $\overline{\varphi(E)}$. Now $\beta \cdot \mathcal{O}_{y,Y}$ is a prime ideal, hence $\mathfrak{P} = \{f \in \mathbb{C}[Y] \,|\, f \in \beta \cdot \mathcal{O}_{y,Y}\}$ is a prime ideal in $\mathbb{C}[Y]$. Moreover $a_1 \notin \mathfrak{P}$ since a_1 and b_1 are relatively prime, hence $a_1 \not\equiv 0$ on $V(\mathfrak{P})$. Therefore, we find

$$Y \supsetneqq V(\mathfrak{P}) \supsetneqq \overline{\varphi(E)}$$

hence $\dim \overline{\varphi(E)} \leq r - 2$.   QED

The stronger version of Zariski's Main Theorem assumes only that $\mathcal{O}_{y,Y}$ is integrally closed in $\mathbb{C}(Y)$. Then either (a) holds, or else the last sentence of (b) holds, i.e., $\varphi^{-1}(y)$ has a positive-dimensional component through y. This is considerably more difficult to prove (cf. Bourbaki, [1], Ch. 5, Ex. 4–7 to §3!). An example where y is not smooth and Zariski's Main Theorem is false will be given below (cf. end of §2B).

§3B. Global Properties—Zariski's Connectedness Theorem, Specialization Principle

We now turn to projective varieties and correspondences $Z \subset X \times Y$. First of all, all the affine results that we have proven apply to the projections

$$p_1 : Z \xrightarrow{\hspace{1.5cm}} X$$
$$p_2 : Z \xrightarrow{\hspace{1.5cm}} Y$$

since these maps are locally projections $\mathbb{C}^{n+m} \to \mathbb{C}^n$. Thus, say $p_1(Z) = X, p_2(Z) = Y$ and assume Z is irreducible. Then:

$$f(x,y) = \max\{\dim W \,|\, W \text{ a component of } p_1^{-1}p_1(x,y) \text{ through } (x,y)\}$$
$$= \max\{\dim W \,|\, W \text{ a component of } Z[x] \text{ through } y\}$$

is an upper semi-continous function *on* Z. And using the fact that $p_1 : Z \to X$ is a closed map, it follows that

$$f(x) = \max\{\dim W \,|\, W \text{ a component of } Z[x]\}$$

is upper semi-continuous on X. Moreover, since $p_1 : Z \to X$ is generically smooth, the value of $f(x)$ almost everywhere is $\dim Z - \dim X$. Therefore we have decomposed X into 2 pieces:

a) a non-empty Zariski-open piece X_0 where all components of $Z[x]$ have dimension $\dim Z - \dim X$.

b) a Zariski-closed piece F where some component of $Z[x]$ has larger dimension: these are called the *fundamental points* of Z.

Note that the codimension in X of all components of F is at least 2. In fact, look at:

$$F^* = \{(x,y) \,|\, f(x,y) > \dim Z - \dim X\} \xrightarrow{\text{res } p_1} F$$
$$\cap$$
$$Z$$

then all components of F^* have dimension at most $\dim Z - 1$ while all fibres of res p_1 have dimension at least $\dim Z - \dim X + 1$, thus all components of F have dimension at most $\dim X - 2$.

Similarly for $Z^{-1} : Y \to X$, we get a non-empty Zariski-open $Y_0 \subset Y$ such that for all $y \in Y_0$, all components of $Z^{-1}[y]$ have dimension $\dim Z - \dim Y$. Therefore, if $x \in X_0$, $y \in Y_0$, we get:

(3.21) $\dim X + \dim(\text{any comp. of } Z[x]) = \dim Y + \dim(\text{any comp. of } Z^{-1}[y])$.

This is known as the "*principle of counting constants*". Now say $\dim X = \dim Z$. Then all the sets $Z[x]$, $x \in X_0$, are finite and by (3.17), it follows that

$$\# Z[x] = [\mathbb{C}(Z):\mathbb{C}(X)],$$

for all x in some smaller Zariski-open set $X_{00} \subset X_0$. In particular, Z is a rational map if and only if $\mathbb{C}(Z) \cong \mathbb{C}(X)$. In other words, when Z is rational, the diagram:

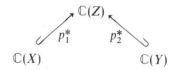

reduces to an injection:

$$\mathbb{C}(X) \xleftarrow{\quad\quad\quad\quad\quad} \mathbb{C}(Y).$$
$$(p_1^*)^{-1} \circ p_2^*$$

(3.22) *I claim that conversely every such injection comes from a unique rational map* Z.

Given any $\alpha: \mathbb{C}(Y) \to \mathbb{C}(X)$, let $Y_0, \ldots, Y_m$ (resp. $X_0, \ldots, X_n$) be homogeneous coordinates on Y (resp. X). Then if $Y_0 \not\equiv 0$ on Y, Y_i/Y_0 generate $\mathbb{C}(Y)$ and if $X_0 \not\equiv 0$, X_i/X_0 generate $\mathbb{C}(X)$. Assuming $Y_0 \not\equiv 0$ on Y, any such α may be described explicitly by

$$\alpha(Y_i/Y_0) = \frac{f_i(X_0, \ldots, X_n)}{f_0(X_0, \ldots, X_n)}$$

where $f_0, \ldots, f_m$ are homogeneous polynomials of the same degree and $f_0 \not\equiv 0$ on X. Let $X_0 = \{x \in X \mid \text{for some } i, f_i(x_0, \ldots, x_n) \neq 0\}$. We can then define a map

$$\phi_\alpha: X_0 \longrightarrow Y$$

by

$$(x_0, \ldots, x_n) \longmapsto (f_0(x), \ldots, f_m(x)).$$

In affine coordinates, in the open set where $x_i \neq 0, f_j(x) \neq 0$, ϕ_α is given by

$$\left(\frac{x_0}{x_i}, \ldots, \frac{x_n}{x_i}\right) \longmapsto \left(\frac{f_0\left(\frac{x_0}{x_i}, \ldots, \frac{x_n}{x_i}\right)}{f_j\left(\frac{x_0}{x_i}, \ldots, \frac{x_n}{x_i}\right)}, \ldots, \frac{f_m\left(\frac{x_0}{x_i}, \ldots, \frac{x_n}{x_i}\right)}{f_j\left(\frac{x_0}{x_i}, \ldots, \frac{x_n}{x_i}\right)}\right),$$

i.e., it is a regular map. Inside $X \times Y$, we have the sets:

$$\left\{\begin{array}{c}\text{Graph of the map} \\ \phi_\alpha: X_0 \to Y\end{array}\right\} \subset V(Y_i f_j(X) - Y_j f_i(X)) \subset X \times Y.$$

call this Z_0 · · · · · · · · · · call this Z^*

Z^* is a closed algebraic set and clearly $Z^* \cap (X_0 \times Y)$ is just Z_0, i.e., the relation from X to Y defined by Z^*, when restricted to Z_0 is just the map ϕ_α. Therefore Z_0 is a closed algebraic subset of $X_0 \times Y$. Note that it is irreducible (in fact, $p_1: Z_0 \to X_0$ is closed and bijective, hence is a homeomorphism and X is irreducible since it is dense in X). Let Z be the Zariski closure of Z_0: equivalently, Z is the unique component of Z^* that is mapped onto X by p_1. We have now constructed a rational map Z from α which is regular on X_0. Note that this Z projects *onto* Y in fact. [For let $g(Y_0, \ldots, Y_m) \equiv 0$ on $p_2(Z)$. Then $g \equiv 0$ on Im ϕ_α, i.e., $g(f_0(X), \ldots, f_m(X)) \equiv 0$ on X_0, hence $\equiv 0$ on X. Then if $d = \deg g$

$$\alpha\left(\frac{g(Y)}{Y_0^d}\right) = \frac{g(f_0(X), \ldots, f_m(X))}{f_0(X)^d} = 0 \text{ in } \mathbb{C}(X)$$

so $g(Y) = 0$ in $\mathbb{C}(Y)$, so $g \equiv 0$ on Y.] We leave it to the reader finally to check that we have just inverted our first construction, so that in fact, there is a bijection between rational maps Z from X onto Y and injections of $\mathbb{C}(Y)$ into $\mathbb{C}(X)$. This is a very fundamental link between geometry and algebra. We have also proved incidentally that every rational map is regular on some non-empty Zariski-open set.

Now consider more closely a general rational map $Z: X \to Y$. We have seen that X has a closed subset F of fundamental points where dim $Z[x] \geq 1$ and that codim $F \geq 2$. On the other hand,

$$X_{\text{reg}} = \{x \in X \mid Z \text{ is regular at } x\}$$

is a non-empty Zariski-open subset of X disjoint from F. In the projective case, Zariski's Main Theorem tells us:

(3.23) **Zariski's Main Theorem.** (Projective-Smooth Case). *Let X and Y be projective varieties and let $Z \subset X \times Y$ be a rational map from X onto Y. Then*

$$Z - F - \text{Sing } X \subset X_{\text{reg}},$$

i.e., *Z is regular at every smooth non-fundamental point.*

Proof. Apply (3.20) to the restriction of $p_1 : Z \to X$ to suitable affines.   QED
As mentioned in §3A, in fact every $x \in X$ which is not fundamental and where $\mathcal{O}_{x,x}$ is integrally closed in $\mathbb{C}(X)$ is actually a regular point. Working globally, Zariski also proved:

(3.24) **Zariski's Connectedness Theorem.** *Let X and Y be projective varieties and let $Z \subset X \times Y$ be a rational map* from X to Y. Then for all topologically unibranch points $x \in X$ (e.g., smooth points), $Z[x]$ is connected in the classical topology, hence also in the Zariski topology.*

Zariski's original theorem assumed about the point x that $\mathcal{O}_{x,x}$ was integrally closed in $\mathbb{C}(X)$ rather than that x was a topologically unibranch point. Now it is true, but hard to prove, that:

$$\mathcal{O}_{x,x} \text{ integrally closed } \Rightarrow X \text{ topologically unibranch at } x.$$

Thus our version really contains the more or less "trivial" part of his original theorem. Moreover, Zariski proved even in char. p, where our definition of topologically unibranch doesn't make sense, that:

$$\mathcal{O}_{x,x} \text{ integrally closed } \Rightarrow Z[x] \text{ connected.}$$

Proof of (3.24). Let $X_0 \subset X$ be a Zariski-open set such that $Z[y] = (\text{one point})$ if $y \in X_0$. Suppose $Z[x] = Y_1 \cup Y_2$, Y_i disjoint classically closed subsets of Y. Let V_i be a classical open neighborhood of Y_i in Y such that $V_1 \cap V_2 = \phi$. Then $Z^{-1}[Y - V_1 - V_2]$ is a classical-closed subset of X not containing x. Since X is topologically unibranch at x, x has a neighborhood U in X such that $X_0 \cap U$ is connected and $Z^{-1}[Y - V_1 - V_2] \cap U = \phi$. This means that $Z[U] \subset V_1 \cup V_2$. But then

$$U \cap X_0 = \{x' \mid Z[x'] \in V_1\} \cup \{x' \mid Z[x'] \in V_2\}$$

and the sets on the right are disjoint and closed in $U \cap X_0$. They are also non-empty because $p_1^{-1}(X - X_0)$ is a proper closed algebraic subset of Z, hence every point $(x,y) \in Z$ is a limit of points $(x_i, y_i) \in Z$ where $x_i \in X_0$. Therefore, for $i = 1, 2$, there is a point $(x', y') \in Z$ with $x' \in U \cap X_0$ and $y' \in V_i$. This shows that $U \cap X_0$ is *not* connected.   QED

We now go on to look at correspondences which are not rational but for which $Z \to X$ is generically finite-to-one.

---

*More generally, the conclusion still holds if, instead of being rational, Z satisfies the condition that $Z[x]$ is Zariski-connected for all x in a Zariski-open $X_0 \subset X$. Namely, by results to be shown in Chapter IV, classically connectedness and Zariski-connectedness are the same, and if $Z[x]$ is classically connected for $x \in X_0$, the same proof goes through.

(3.25) **Specialization Principle.** *Let $Z \subset \mathbb{P}^n \times \mathbb{P}^m$ be an r-dimensional subvariety and let $X = p_1(Z) \subset \mathbb{P}^n$. Suppose* $\dim X = r$ *too so that*

$$\varphi = \operatorname{res} p_1 : Z \to X.$$

is almost everywhere finite-to-one. Let $d = [\mathbb{C}(Z):\mathbb{C}(X)]$. If $x \in X$ is topologically unibranch, and $\varphi^{-1}(x)$ is a finite set then

$$\sum_{\phi(y)=x} \operatorname{mult}_y(\varphi) = d.$$

Moreover, the sets

$$X_1 = \{x \in X \mid X \text{ smooth at } x \text{ and } \varphi^{-1}(x) \text{ finite}\}$$
$$X_0 = \{x \in X_1 \mid \varphi \text{ is smooth at all } y \in \varphi^{-1}x\}$$

are Zariski-open and $x \longrightarrow \# \varphi^{-1}(x)$ is lower semi-continuous in the Zariski topology on X_1 and constant* with value d on X_0. d is called the degree of Z or deg Z.

   Proof. By (3.16), $Z_1 = \{z \in Z \mid \{z\} \text{ is a component of } \varphi^{-1}\varphi(z)\}$ is Zariski-open in Z. Therefore $\varphi(Z - Z_1) \cup \operatorname{Sing} X$ is closed. By definition, this set is the complement of X_1 so X_1 is Zariski-open. Similarly $Z_0 = \{z \in Z_1 \mid z, \varphi(z) \text{ are smooth and } \varphi \text{ is smooth at } z\}$ is Zariski-open in Z, and since $X - X_0 = \varphi(Z - Z_0)$, so is X_0. Next by (3.17), $\# \varphi^{-1}(y) = d$ on some Zariski-open $X_0' \subset X_0$. Moreover by (3.17) plus a straightforward induction on $\dim X$, we see that $x \mapsto \# \varphi^{-1}(x)$ is a *cons-tructible* function on X_1, i.e., $\{x \mid \# \varphi^{-1}(x) = n\}$ is a constructible set.
   Next we apply (3.10): if $x \in X$ is a topologically unibranch point where $\varphi^{-1}(x) = \{y_1, \ldots, y_t\}$ is finite, choose disjoint neighborhoods V_i of y_i and a neighborhood U of x such that:
   a) $\varphi^{-1}U = V_1 \cup \ldots \cup V_t$,
   b) $U \cap X_0'$ is connected.
Then by Definition (3.12),

$$\sum_{i=1}^{t} \operatorname{mult}_{y_i}(\varphi) = \sum_{i=1}^{n} \# (\varphi^{-1}(z) \cap V_i)$$
$$= \# \varphi^{-1}(z)$$
$$= d$$

for any $z \in U \cap X_0'$. Moreover φ maps each V_i onto U (as in proof of (3.10)), so for each $z \in U$, $\# \varphi^{-1}(z) \geq t$. This proves that $x \to \# \varphi^{-1}(x)$ is lower semi-continuous in the classical topology. Added to the fact that $\# \varphi^{-1}(x)$ is a constructible function, $\# \varphi^{-1}(x)$ must be lower semi-continuous in the Zariski topology too.
   Actually the form of the Specialization Principle that we want to apply in Ch. V is slightly more general in that the irreducibility of Z is almost irrelevant. What we will use is:

(3.26) **Specialization Principle** (2nd form). *Let $Z \subset \mathbb{P}^n \times \mathbb{P}^m$ be a closed algebraic set and let $X = p_1(Z)$. Assume X irreducible of dimension r. Let $Z = Z_1 \cup \ldots \cup Z_k$,*

---

*Infinitively, the idea is that $\operatorname{res}(\varphi): \varphi^{-1}(X_0) \to X_0$ is a covering map with d sheets, but that over $X_1 - X_0, \varphi$ ramifies, various sheets come together and $\# \varphi^{-1}(x)$ decreases: e.g. consider the map $\varphi: \mathbb{P}^1 \to \mathbb{P}^1$ given by $\varphi(t) = t^2, \varphi(\infty) = \infty$.

let $\varphi = \mathrm{res}\ p_1 : Z \to X$ *and let* $\varphi_i = \mathrm{res}\ p_1 : Z_i \to X$. *If* $x \in X$ *is topologically uni-branch,* $\varphi^{-1}(x)$ *is a finite set, and* $\varphi^{-1}(x)$ *does not meet any components* Z_i *of* Z *of dimension* $< r$, *then*

$$\sum_{\phi(y)=x} \mathrm{mult}_y(\varphi) = d$$

where now

$$\mathrm{mult}_y(\varphi) = \sum_{\left[\begin{smallmatrix} \text{all } i \text{ for} \\ \text{which } y \in Z_i \end{smallmatrix}\right]} \mathrm{mult}_y(\varphi_i)$$

$$d = \sum_{\left[\begin{smallmatrix} \text{all } i \text{ for} \\ \text{which } Z_i \to X \\ \text{is surjective} \end{smallmatrix}\right]} [\mathbb{C}(Z_i) : \mathbb{C}(X)]$$

Moreover, the sets

$$X_1 = \left\{ x \in X \,\middle|\, \begin{matrix} X \text{ smooth at } x,\ \varphi^{-1}(x) \text{ finite, all } Z_i \text{ meeting} \\ \varphi^{-1}(x) \text{ have dim} \geq r, \text{ hence dim} = r \text{ by } (3.13) \end{matrix} \right\}$$

$$X_0 = \left\{ x \in X_1 \,\middle|\, \begin{matrix} \text{each } z \in \varphi^{-1}(y) \text{ is on a unique } Z_i \text{ and} \\ \text{this } Z_i \text{ is smooth at } z \text{ over } X \end{matrix} \right\}$$

are Zariski-open and $x \mapsto \#\ \varphi^{-1}(x)$ *is lower semi-continuous on* X_1, *constant with value* d *on* X_0.

(Proof left to reader.)

An interesting example of the Specialization Principle is the case where $X \subset \mathbb{P}^n$ is a smooth curve, $f \in \mathbb{C}(X)$, and f defines a rational map

$$Z : X \longrightarrow \mathbb{P}^1$$

as in (2.17). In fact, by Zariski's Main Theorem (since X is smooth and dim $Z =$ dim $X = 1$, hence $Z[x] \cong Z \cap (\{x\} \times \mathbb{P}^1) \subsetneqq Z$ must be finite), Z is regular, and we have:

$$\begin{array}{ccc} & Z & \\ {}^{\approx}\swarrow {\scriptstyle p_1} & & {\scriptstyle p_2}\searrow \\ X & & \mathbb{P}^1. \end{array}$$

We apply the specialization principle to p_2. Note that for all $a \in \mathbb{P}^1$,

$$p_2^{-1}(a) \cong Z^{-1}[a] = \{x \in X \,|\, f(x) = a\}$$

(here $f(x) = \infty$ is to mean that x is a pole of f: cf. (2.18)).

Note also that $\mathbb{C}(Z) \cong \mathbb{C}(X)$ and if t is the coordinate on $\mathbb{P}^1$, then $Z^*(t) = f$. Therefore:

$$[\mathbb{C}(Z) : \mathbb{C}(\mathbb{P}^1)] = [\mathbb{C}(X) : \mathbb{C}(f)].$$

Call this number d. The specialization principle now says that

$$\#\ \{x \in X \,|\, f(x) = a\}$$

equals d for all but a finite set of values a, and is less than d for these special values. Moreover, for all a we can define $f^{-1}(a)$ *as a divisor on* X by

$$f^{-1}(a) = \sum_{\substack{x \text{ such that} \\ f(x)=a}} \text{mult}_x(Z) \cdot x$$

and then for all a:

$$\text{degree}\,(f^{-1}(a)) = d$$

where degree is the unique linear function on $\text{Div}(X)$ assigning degree one to each point.

This is a little confusing because we also have an alternate definition of $f^{-1}(a)$ as a pull-back by the regular map Z of the divisor on $\mathbb{P}^1$ consisting of the one point a (cf. discussion following (2.22)); and if $a = 0$ or ∞, we have a third definition of $f^{-1}(a)$ as the divisor of zeroes or of poles of f (cf. (2.17)). Actually as they should, all these definitions come out the same.

Thus if $x \in X$, and $\xi \in \mathfrak{M}_{x,X}$ is a local equation of the divisor x, then ξ generates $\mathfrak{M}_{x,X}, d\xi$ is non-zero on $T_{x,X}$ and ξ is a local analytic coordinate at x. Say $f(x) = a$ where $a \neq \infty$. Then the multiplicities of a in $f^{-1}(a)$ as a pull-back of divisors is gotten by taking the local equation t-a of a as a divisor on $\mathbb{P}^1$, pulling it back to $f - a$, writing

$$f - a = u \cdot \xi^k, u \text{ unit in } \mathcal{O}_{x,X}$$

and then $k \cdot x$ is $f^{-1}(a)$ locally near x. But ξ defines an analytic isomorphism of a neighborhood $U \subset X$ of x with a small disc Δ around 0 in the complex ξ-plane. And for a' near a,

$$f^{-1}(a') \cap U \cong \{\xi \in \Delta \mid u \cdot \xi^k = a' - a\}.$$

As the function $u\xi^k$ has a k-fold zero at $\xi = 0$, for all small $a' - a$, $u \cdot \xi^k = a' - a$ has k solutions near 0. Thus k is also $\text{mult}_x Z$. If $a = \infty$, a similar argument holds. Also if $a = 0$ or ∞, it is really only a restatement of the definition to see that divisor of zeroes (resp. poles) of f equals "divisor-pull-back" $f^{-1}(0)$ (resp. $f^{-1}(\infty)$).

A consequence of this computation is the important Corollary:

(3.27) **Corollary.** *If X is a smooth curve, $f \in \mathbb{C}(X)$, and degree is linear function on $\text{Div}(X)$, equal to one on each point, then* degree *(zeroes of f) =* degree *(poles of f), hence degree is a function on $\text{Div}(X)$ mod principal divisors (f), i.e. $\text{Pic}(X)$.*

At several points in this chapter, the assumption of topological unibranch has come up. We want to describe a famous example of a correspondence between varieties one of which is *not* topologically unibranch which shows that in all these cases, this hypothesis was necessary. Let $C \subset \mathbb{P}^2$ be the plane curve defined by:

$$XYZ = X^3 + Y^3$$

and let C_0 be the affine piece $Z \neq 0$, defined in affine coordinates by

$$xy = x^3 + y^3.$$

Near $(0,0)$, C_0 is very closely approximated by the reducible curve $xy = 0$. In fact if

$$\max(|x|, |y|) = \varepsilon,$$

then

$$|x^3 + y^3| \leq 2\varepsilon^3$$

so either

$$|x| = \varepsilon, |y| \leq 2\varepsilon^2$$

or

$$|x| \leq 2\varepsilon^2, |y| = \varepsilon.$$

The inequalities $|y| \leq 2|x|^2$ and $|x| \leq 2|y|^2$ define 2 branches of C_0 at $(0,0)$ and C_0 is *not* topologically unibranch. Consider the everywhere single-valued rational map

$$Z : \mathbb{P}^1 \to C$$

defined in homogeneous coordinates by

$$Z(S, T) = (ST^2, S^2T, S^3 + T^3).$$

Z restricts to the regular map

$$Z_0 : \mathbb{C} - \{-1, -e^{2\pi i/3}, -e^{4\pi i/3}\} \to C_0$$

given by

$$Z_0(s) = \left(\frac{s}{1 + s^3}, \frac{s^2}{1 + s^3} \right)$$

and, in fact, it is not hard to check that Z is everywhere regular (this also follows by Zariski's Main Theorem). Noting that under Z,

$$Z^*(y/x) = S/T$$

it is not hard to check that for all $P \in C$ *except* $(0,0,1)$, $Z^{-1}(P)$ consists of one point, while

$$Z^{-1}(0,0,1) = \{(0,1),(1,0)\}.$$

This example now illustrates many phenomena:

　　i) at $(0,1)$, Z_0 is not open because Z_0 (neigh of $(0,1)$) contains only the branch of C_0 defined by $|y| \leq 2|x|^2$. Thus (3.10) is false if assumption (a) is dropped.

　　ii) $Z^{-1}(0,0,1)$ is a finite set, hence $(0,0,1)$ is not a fundamental point of Z^{-1}, yet Z^{-1} is not regular at $(0,0,1)$. Thus Zariski's Main Theorems (3.20) and (3.23) are false if the hypothesis of smoothness is dropped,

　　iii) $Z^{-1}(0,0,1)$ is disconnected so Zariski's connectedness theorem (3.24) is false if the top. unibranch hypothesis is dropped,

　　iv) Z is birational, hence has degree 1, so under no assignments of positive multiplicities can we have

$$\sum_{Z(y)=(0,0,1)} \text{mult}_y(Z) = \text{degree}.$$

Thus the specialization principle (3.25) requires the topologically unibranch hypothesis.

§3C. Intersections on Smooth Varieties

As a final topic in this section, I want to go back to Corollary (3.14) and study its generalizations to the dimension of the intersection of two subvarieties of a variety. The main result is this:

(3.28) **Proposition.** *Let Z^n be an affine variety and let $X^r, Y^s \subset Z^n$ be subvarieties (superscript indicates dimension). Let $x \in X \cap Y$ and assume x is smooth on Z. Then if*

$$X \cap Y = W_1 \cup \ldots \cup W_k \cup (\text{components not through } x)$$
$$x \in W_i, \quad 1 \leq i \leq k$$

then

$$\dim W_i \geq r + s - n.$$

Proof. We prove this by a very old idea of not thinking of a variable point on Z that must be simultaneously on X and Y; but rather of 2 variable points, one on X, one on Y that must be made equal! Precisely, look at the diagonal map

$$\delta : Z \longrightarrow Z \times Z$$

and let $\Delta = \delta(Z)$ be the diagonal. Then $\delta(X \cap Y) = (X \times Y) \cap \Delta$. Since $\dim(X \times Y) = r + s$, the result follows immediately from (3.14) if we prove:

(3.29) **Lemma.** *If x is a smooth point of Z^n, then $\exists$ functions $f_1, \ldots, f_n$ on $Z \times Z$ such that $V(f_1, \ldots, f_n) = \Delta \cup (\text{components not thru } (x,x))$.*

Proof. In fact, let $g_1, \ldots, g_n$ be functions on Z, zero at x with independent differentials at x. Then let $f_i(X, X') = g_i(X) - g_i(X')$ be functions on $Z \times Z$ (X, X' being coordinates in the 2 factors). Clearly $f_1, \ldots, f_n$ are zero on Δ and have independent differentials at (x,x). Therefore by (1.16), $V(f_1, \ldots, f_n)$ has a unique n-dimensional component through (x,x) and this must be Δ. QED

(3.30) **Corollary.** *Let $X^r, Y^s \subset \mathbb{P}^n$ be subvarieties such that $r + s \geq n$. Then $X \cap Y \neq \phi$.*

Proof. If $X = V(f_i)$, $Y = V(g_j)$, where $f_i, g_j \in \mathbb{C}[X_0, \ldots, X_n]$ are homogeneous, define X^* and Y^* to be the zeroes of the f_i's and g_j's in affine space $\mathbb{C}^{n+1}$. If

$$\pi : \mathbb{C}^{n+1} - (0, \ldots, 0) \longrightarrow \mathbb{P}^n$$

is the map taking a point $(a_0, a_1, \ldots, a_n)$ to the point in projective space with these as homogeneous coordinates, then $X^* = \pi^{-1}(X) \cup \{0\}$ and $Y^* = \pi^{-1}(Y) \cup \{0\}$. Now

$$\mathbb{C}(X^*) = \begin{cases} \text{field of } f/g\text{'s}, f, g \in \mathbb{C}[X_0, \ldots, X_n], \\ g \not\equiv 0 \text{ on } X \text{ modulo those with } f \equiv 0 \text{ on } X \end{cases}$$

contains $\mathbb{C}(X)$, the field of f/g's, with f, g homogeneous of same degree, as a subfield. Moreover, if $X_0 \not\equiv 0$ on X^*, then

$$\mathbb{C}(X^*) = \mathbb{C}(X)(X_0).$$

It is easy to verify that X_0 is a transcendental over $\mathbb{C}(X)$: hence

$$\dim X^* = \text{tr.d. } \mathbb{C}(X^*) = 1 + \text{tr.d. } \mathbb{C}(X) = 1 + \dim X = 1 + r.$$

Similarly $\dim Y^* = 1 + s$. But now $0 \in X^* \cap Y^*$. By (3.28), $\dim X^* \cap Y^* \geq (r+1) + (s+1) - (n+1) \geq 1$. Therefore $X^* \cap Y^*$ contains more points! Applying π, we find points in $X \cap Y$. QED

Here is an example to illustrate that (3.28) is false if you drop the assumption that x is smooth on Z:

Take $Z^3 = $ (affine variety $xy = uv$) in $\mathbb{C}^4$
Take $X^2 = $ (plane $x = u = 0$) $\subset Z^3$
Take $Y^2 = $ (plane $y = v = 0$) $\subset Z^3$
Then $X^2 \cap Y^2 = $ origin only

i.e., $\dim(X \cap Y) = 0 < 2 + 2 - 3 = \dim X + \dim Y - \dim Z$.

Chapter 4. Chow's Theorem

§4A. Internally and Externally Defined Analytic Sets and their Local Descriptions as Branched Coverings of $\mathbb{C}^n$

So far we have used complex analytic methods as a tool in our study of algebraic varieties. It is natural to ask to what extent the general theory of complex analytic spaces parallels that of algebraic varieties. The answer seems to be that there is a very close parallel, at least in the local theory, except that everything is a bit harder to prove, due to the fact that analytic functions, unlike rational functions, may have "essential singularities". I do not want to go into this in detail as there are excellent references (Gunning-Rossi [1], Hervé [1], Gunning [1]) but will merely describe some of the parallels and prove one result on when "essential singularities" do *not* occur. This result has a consequence, the famous theorem of Chow: that the only complex analytic subsets of $\mathbb{P}^n$ are algebraic varieties. This should be viewed as a generalization of the old result that the only everywhere meromorphic functions on $\mathbb{C} \cup \{\infty\}$ are rational functions. It is one of the keys in linking analytic and algebraic geometry.

First some general definitions:

(4.1) Definition. *Let $U \subset \mathbb{C}^n$ be an open set. A closed subset $X \subset U$ is an analytic subset of U if for all $x \in X$, there is an open neighborhood $U' \subset U$ of x and a finite set of analytic functions $f_1, \dots, f_k$ defined on U' such that*

$$X \cap U' = \{y \in U' | f_1(y) = \dots = f_k(y) = 0\}.$$

Variants of this are: 1) if $x_0 \in U$ is fixed and U is allowed to shrink to any smaller neighborhood of x_0, then we obtain the *germ* of an analytic subset of $\mathbb{C}^n$ at x_0. Or, 2) if $X \subset \mathbb{P}^n$ is a closed subset such that for every $x \in X$, X is defined near x by a finite set of analytic functions in affine coordinates at x, then X is an analytic subset of $\mathbb{P}^n$. Also 3) $X \subset U$ is said to be an *analytic submanifold* at x if X is defined near x by k functions $f_1, \dots, f_k$ with independent differentials at x: then by the implicit function theorem, X is an $(n - k)$-dimensional complex manifold near x, and 4) $X \subset U$ is said to be *irreducible* if X cannot be decomposed into $X_1 \cup X_2$, where X_i are smaller analytic subsets of U. Some other fundamental definitions are:

$$\mathbb{C}\{X_1, \dots, X_n\} = \text{ring of convergent power series, i.e., germs of analytic functions at } 0 \in \mathbb{C}^n$$

and for all analytic sets $X \subset U$, and $x \in X$,

$$\mathcal{O}^{an}_{x,X} = \text{ring of germs of functions on } X \text{ at } x \text{ which are restrictions of analytic functions on } U$$
$$\cong \mathbb{C}\{X_1, \dots, X_n\} / \mathfrak{A}$$

(if $\mathfrak{A}$ = ideal of functions, which when translated from 0 to x, vanish on X). A holomorphic, or analytic map between analytic sets $X_1 \subset U_1 \subset \mathbb{C}^{n_1}$ and $X_2 \subset U_2 \subset \mathbb{C}^{n_2}$ is a map $f : X_1 \to X_2$ given locally by expressing the coordinates $Y_1, \dots, Y_{n_2}$ in $\mathbb{C}^{n_2}$ as analytic functions of the coordinates $X_1, \dots, X_{n_1}$ in $\mathbb{C}^{n_1}$. If 2 analytic sets X_1, X_2 are isomorphic, i.e., there are inverse holomorphic maps $f : X_1 \to X_2, f^{-1} : X_2 \to X_1$, then the local rings $\mathcal{O}^{an}_{x_1, X_1}$ and $\mathcal{O}^{an}_{x_1, X_2}$ at corresponding points are isomorphic.

(4.1) is the "external" characterization of analytic sets, i.e., defining them by cutting down from $\mathbb{C}^n$. The other approach is to build them up "internally" as unions of submanifolds and singular limit points. From this point of view, one would define:

(4.2) **Definition.** *Let $U \subset \mathbb{C}^n$ be an open set. A closed subset $X \subset U$ is a *-analytic subset of U if X can be decomposed:*

$$X = X^{(r)} \cup X^{(r-1)} \cup \dots \cup X^{(0)},$$

where for all i, $X^{(i)}$ is an i-dimensional complex submanifold of U and $\overline{X^{(i)}} \subset X^{(i)} \cup X^{(i-1)} \cup \dots \cup X^{(0)}$. If $X^{(r)} \neq \phi$, then r is called the dimension of X.

In fact, this definition is equivalent to (4.1): X is analytic $\iff$ X is *-analytic. The fact that analytic implies *-analytic is part of the standard local theory of analytic sets. In fact, one shows:

(4.3) **Proposition.** i) *Every analytic X can be decomposed uniquely: $X = \bigcup_{i \in I} X_i$, where X_i are irreducible, the union is locally finite and $X_i \nsubseteq X_j$, any i, j.*

ii) *Every irreducible X is the union of an open, dense r-dimensional complex manifold U, and a closed analytic subset $\text{Sing } X$ of points $x \in X$ where X is not a manifold. r is called the dimension of X. Every component of $\text{Sing } X$ has dimension less than r.*

For proofs of this, see Gunning-Rossi, [1], pp. 116, 141, 155, or Hervé, [1], Ch. IV, Th. 6 and 8. Using (4.3), we immediately see:

(4.4) **Corollary.** *An analytic set X is *-analytic.*

Proof. In fact, decompose X as in (i), let $r = \max \dim X_i$, and set

$$X^{(r)} = X - \bigcup_{i \in I} \text{Sing } X_i - \bigcup_{\dim X_i < r} X_i.$$

Then $X' = X - X^{(r)}$ is a closed analytic subset of X of lower dimension and repeat the procedure.   QED

However, we will avoid using (4.3) (except in the appendix to Chapter 6). In fact, for us, the interesting result is the converse:

(4.5) **Theorem.** *A *-analytic set is analytic.*

This may be thought of as saying that if an r-dimensional submanifold $X^{(r)} \subset U$ has its closure contained in a union of smaller dimensional submanifolds X', then $X^{(r)} \cup X'$ is an analytic set. In fact the closure $\overline{X^{(r)}}$ itself is analytic. Note that this is false if $\dim X'$ isn't smaller than $\dim X^{(r)}$: look at

$$X^{(1)} = \text{locus of points } (x, e^{1/x}) \text{ in } \mathbb{C}^2, x \in \mathbb{C} - (0)$$
$$X' = \text{locus of points } (0, y) \text{ in } \mathbb{C}^2, y \in \mathbb{C}.$$

Then $X^{(1)}$ is an analytic submanifold of $\mathbb{C}^2 - X'$, X' is an analytic submanifold of $\mathbb{C}^2$ and $\bar{X}^{(1)} = X^{(1)} \cup X'$. But $X^{(1)} \cup X'$ is *not* analytic.

Before beginning to prove (4.5), note how Chow's theorem is a Corollary:

(4.6) **Corollary.** *If* $X \subset \mathbb{P}^n$ *is a closed *-analytic subset, then* X *is a finite union of algebraic varieties.*

Proof. Consider the natural map

$$\pi : \mathbb{C}^{n+1} - (0) \longrightarrow \mathbb{P}^n$$

and let $CX = \pi^{-1}(X)$ be the "cone" over X. Then CX is *-analytic, but with strata of one higher dimension than those of X. Now in $\mathbb{C}^{n+1}$, consider $Z = CX \cup \{0\}$. This is closed and is *-analytic with $\{0\} = Z^{(0)}$ being the zero-dimensional strata. Therefore by (4.5), Z is analytic. Let $f_1, \ldots, f_k$ be analytic functions in a neighborhood of 0 such that Z is defined by $f_1 = \ldots = f_k = 0$. Note that Z is invariant under scalar multiplication $\bar{x} \to \lambda \cdot \bar{x}$, hence for all $\lambda \in \mathbb{C}$, $f_i^\lambda(x_0, \ldots, x_n) = f_i(\lambda x_0, \ldots, \lambda x_n)$ is also zero on Z. Now write out

$$f_i(x) = \sum_\alpha c_\alpha^{(i)} \cdot x^\alpha$$

$\alpha = (\alpha_0, \ldots, \alpha_n)$ being a multi-index.

Let

$$f_{i,r}(x) = \sum_{|\alpha| = r} c_\alpha^{(i)} x^\alpha.$$

Then

$$f_i^\lambda(x) = \sum_{r=0}^\infty \lambda^r \cdot f_{i,r}(x)$$

vanishes on Z for all λ. Therefore

$$\left(\frac{d}{d\lambda} \right)^s f_i^\lambda(x) = \sum_{r=s}^\infty \binom{r}{s} \lambda^{r-s} \cdot f_{i,r}(x)$$

vanishes on Z for all s and λ. Setting $\lambda = 0$, it follows that $f_{i,r}(x)$ vanishes on Z, i.e., Z is the set of zeroes of the *homogeneous polynomials* $f_{i,r}, 1 \leq i \leq k, 1 \leq r < \infty$. It follows that X is the algebraic set $V(\ldots, f_{i,r}, \ldots)$ (by Hilbert's Basis Theorem, the infinite set $\{f_{i,r}\}$ may be replaced by a finite subset), hence X is a finite union of algebraic varieties. QED

The proof of (4.5) is based on the familiar method of representation of an analytic or algebraic set locally as a branched covering by means of a projection. There will be many parallels with Chapters 2 and 3. We use induction on the dimension of the *-analytic set X being studied. Thus it will suffice to prove the following:

(4.7) **Induction step.** *Given* $X = X^{(r)} \cup X' \subset U \subset \mathbb{C}^n$

$\qquad$ *where* $\quad X^{(r)} = r$-*dimensional manifold*

$\qquad\qquad\qquad X' \; = $ *analytic and* *-*analytic set of* dim. $< r$

$\qquad$ *assume* $\quad \overline{X^{(r)}} \subset X^{(r)} \cup X'$

$\qquad\qquad$ *then* $\quad \overline{X^{(r)}}$ *is an analytic set.*

For technique, we will need in our proof 2 very fundamental facts about analytic functions. The first is:

(4.8) **Weierstrass Preparation Theorem.** *Let* $f \in \mathbb{C}\{X_1,\dots,X_n\}$ *and assume*

$$f(0,\dots,0,X_n) = \alpha \cdot X_n^d + \text{higher terms}, \alpha \neq 0.$$

Then there are unique functions $u \in \mathbb{C}\{X_1,\dots,X_n\}$ *with* $u(0) \neq 0$, *and* $a_i \in \mathbb{C}\{X_1,\dots, X_{n-1}\}$, $1 \leq i \leq d$ *with* $a_i(0) = 0$, *such that:*

$$f = u \cdot (X_n^d + a_1 X_n^{d-1} + \dots + a_d),$$

i.e., f *differs by a unit from an element of* $\mathbb{C}\{X_1,\dots,X_{n-1}\}[X_n]$. *Moreover, for all* $g \in \mathbb{C}\{X_1,\dots,X_n\}$, *there are unique functions* $h \in \mathbb{C}\{X_1,\dots,X_n\}$ *and* $b_i \in \mathbb{C}\{X_1,\dots, X_{n-1}\}$, $1 \leq i \leq d$, *such that:*

$$g = h \cdot f + (b_1 X_n^{d-1} + \dots + b_d).$$

A proof may be found in $Z - S$, vol. 2, pp. 139–149, Gunning-Rossi, [1], pp. 67–72 or Hervé, [1], pp. 10–12. One of the standard Corollaries of (4.8) is:

(4.9) **Corollary.** $\mathbb{C}\{X_1,\dots,X_n\}$ *is a UFD.*

The second fact is:

(4.10) **Riemann Extension Theorem.** *Let* $U \subset \mathbb{C}^n$ *be open and let* $X \subset U$ *be an analytic set. Let* f *be an analytic function on* $U - X$ *such that for all* $x \in X$, f *is bounded in some neighborhood of* x. *Then* f *extends to an analytic function on* U.

This is easily proven*, using (4.8) and the even more basic fact (which may even be taken as definition) that analytic functions are characterized as C^1-

---

*The proof goes as follows: take any point $x \in X$ which, by changing coordinates, we assume to be origin, and let g be a non-zero analytic function, defined near x and zero on X. By (4.8) we may assume $g \in \mathbb{C}\{X_1,\dots,X_{n-1}\}[X_n]$ and is monic in X_n of some degree d and by (4.9), we may assume g has no multiple factors when factored in the UFD $\mathbb{C}\{X_1,\dots,X_{n-1}\}[X_n]$ (otherwise replace g by the product of its irreducible factors). Since $\mathbb{C}\{X_1,\dots,X_{n-1}\}$ has characteristic O, this implies that the discriminant δ of g, considered as a polynomial in X_n, is not zero. Note that as δ is a polynomial in the coefficients of f, $\delta \in \mathbb{C}\{X_1,\dots,X_{n-1}\}$. Let X' be the locus of zeroes of g. Now look at the functions $g.f$ and $g^2 \cdot f$ defined near x. $g \cdot f$ is continuous and zero on X', and $g^2 f$ is even in C^1 with derivatives zero on X'. Since $g^2 f$ is analytic off X, it satisfies the Cauchy-Riemann equations off X, and since its derivatives are zero on X', it also satisfies the Cauchy-Riemann equations on X'. Thus $g^2 f \in \mathbb{C}\{X_1,\dots,X_n\}$. To finish the proof, we need only check:

(*) $\qquad\qquad\qquad h \in \mathbb{C}\{X_1,\dots,X_n\}, h$ zero on $X' \Rightarrow h/g \in \mathbb{C}\{X_1,\dots,X_n\}$.

By (4.8) again, in the proof of (*) we may assume h is a polynomial in X_n of degree less than d. Now for almost all small $a_1,\dots,a_{n-1}, \delta(a_1,\dots,a_{n-1}) \neq 0$, hence the polynomial $h(a_1,\dots,a_{n-1},X_n)$ has d distinct roots $a_n^{(i)}$. All the points $(a_1,\dots,a_{n-1},a_n^{(i)})$ are in X', hence h is zero on them. This gives $h(a_1,\dots,a_{n-1},X_n)$ d distinct roots. Thus the coefficients of h are zero at all such $(a_1,\dots,a_{n-1})$. Since these are dense, $h \equiv 0$ and (*) is proven.

complex-valued functions f which satisfy the Cauchy-Riemann equations:

(4.10)
$$\frac{\partial f}{\partial u_i} = -i\,\frac{\partial f}{\partial v_i}, \qquad 1 \le i \le n$$

where u_i, v_i are the real and imaginary parts of the complex coordinates $X_1,\ldots,X_n$.

The first step in the proof is to analyze the effect of projections on analytic sets. The most elementary approach is to use Weierstrass Preparation plus resultants as follows:

(4.11) Proposition (Elimination theory for analytic sets). *Let $p:\mathbb{C}^{n+k} \to \mathbb{C}^n$ be a linear projection, let $U \subset \mathbb{C}^{n+k}$, $V \subset \mathbb{C}^n$ be open sets with $p(U) \subset V$ and let $X \subset U$ be an analytic set. Assume res $p:X \to V$ is proper. Then $p(X)$ is an analytic set in V and res $p:X \to V$ is a finite-to-one map.*

Proof. By induction on k, we readily reduce the general case to the special case $p:\mathbb{C}^{n+1} \to \mathbb{C}^n$: in fact, factor p:

$$\mathbb{C}^{n+k} \xrightarrow{\;\;p_1\;\;} \mathbb{C}^{n+k-1} \xrightarrow{\;\;p_2\;\;} \mathbb{C}^n$$

and let $U_1 = p_1(U) \subset \mathbb{C}^{n+k-1}$. Consider the maps

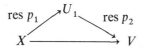

It follows that res p_1 is proper, hence by the special case $X_1 = p_1(X)$ is analytic and res p_1 is finite-to-one. But then one sees that res $p_2:X_1 \to V$ is proper so by induction $p_2(X_1) = p(X)$ is analytic and $\mathrm{res}_{X_1} p_2$ is finite-to-one, hence res p is finite-to-one.

Now say $k = 1$ and take any point $y \in V$. Then $X \cap p^{-1}(y)$ is an analytic subset of $U \cap p^{-1}(y)$, i.e., of an open subset of $\mathbb{C}$; and $X \cap p^{-1}(y)$ is compact. Therefore $X \cap p^{-1}(y)$ is a finite set of points and res p is finite-to-one. Suppose $X \cap p^{-1}(y) = \{x_1,\ldots,x_k\}$. Let U_i be disjoint neighborhoods of the points x_i. Then if $V_1 \subset V$ is a small enough neighborhood of y, $X \cap p^{-1}(V_1) \subset U_1 \cup \ldots \cup U_k$, hence $X \cap p^{-1}(V_1)$ breaks up into disjoint open subsets $X_i = X \cap p^{-1}(V_1) \cap U_i$. Thus $p(X)$, near y, is the union $p(X_1) \cup \ldots \cup p(X_k)$ and it suffices to prove that $p(X_i)$ is an analytic subset of V near y. In fact, replacing V by V_1, U by $U_i \cap p^{-1}(V_1)$, X by X_i, we are reduced to the case $p^{-1}(y)$ consists in one point. For convenience, change coordinates now so that $y = 0$, $p^{-1}(y) = \{0\}$.

Now say X is defined in a neighborhood $U_1 \subset U$ of 0 by analytic functions $f_1,\ldots f_k \in \mathbb{C}\{X_1,\ldots,X_{n+1}\}$. Shrinking U_1 if necessary and applying Weierstrass Preparation (4.8), we can assume

$$f_1 = X_{n+1}^d + a_1 X_{n+1}^{d-1} + \ldots\ldots + a_d$$
$$f_i = \qquad\quad b_{1i} X_{n+1}^{d-1} + \ldots\ldots + b_{di}, \, 2 \le i \le k$$
$$a_i, b_{ij} \in \mathbb{C}\{X_1,\ldots,X_n\}, a_i(0) = 0.$$

Considering f_1 and $\sum_{i=2}^{k} t_i f_i (t_i \in \mathbb{C})$ as polynomials in X_{n+1}, form their resultant. This will have the form:

$$R(f_1, t_2 f_2 + \ldots + t_k f_k) = \sum_{|\alpha|=d} t^\alpha \cdot R_\alpha, \quad R_\alpha \in \mathbb{C}\{X_1, \ldots, X_n\}.$$

Let $V_1 \subset V$ be a neighborhood of 0 small enough so that

    a) all the a_i, b_{ij} and hence the R_α converge in V_1,

    b) if $(x_1, \ldots, x_n) \in V_1$, then all roots of $f_1(x_1, \ldots, x_n, X_{n+1}) = 0$ are in U_1.

Let $U' = U_1 \cap p^{-1}(V_1)$. Then I claim $p(X \cap U')$ is the set of common zeroes in V_1 of the R_α's. In fact, if $(x_1, \ldots, x_{n+1}) \in X \cap U'$, then for all $t_i, f_1(x_1, \ldots, x_n, X_{n+1})$ and $\sum_{i=2}^{k} t_i f_i(x_1, \ldots, x_n, X_{n+1})$ have the common root $X_{n+1} = x_{n+1}$, so $\sum t^\alpha R_\alpha(x_1, \ldots, x_n) = 0$. Since this is true for all t's, $R_\alpha(x_1, \ldots, x_n) = 0$ for all α. Conversely, if $(x_1, \ldots, x_n) \in V_1$ and $R_\alpha(x_1, \ldots, x_n) = 0$, all α, then $f_1(x_1, \ldots, x_n, X_{n+1})$ and $\sum_{i=2}^{k} t_i \cdot f_i(x_1, \ldots, X_n, X_{n+1})$ have a common root for all t. Let $\alpha_1, \ldots, \alpha_d$ be the roots of the 1st equation and let $W_j \subset \mathbb{C}^{k-1}$ be the set of all $(t_2, \ldots, t_k)$ such that $\sum_{i=2}^{k} t_i f_i(x_1, \ldots, x_n, \alpha_j) = 0$. Then W_j is a subvector space and we have seen that $W_1 \cup \ldots \cup W_d = \mathbb{C}^{k-1}$. Therefore for some j_0, $W_{j_0} = \mathbb{C}^{k-1}$, hence $(x_1, \ldots, x_n, \alpha_{j_0})$ is a root of all the f_i's, hence is in $X \cap U'$.   QED

However, we are in a slightly messier situation: $X_0 \cup X_1 \subset U$ where X_1 is closed and analytic in U and X_0 is closed and analytic in $U - X_1$. If we try to project such a set, we first need:

(4.12) Lemma. *Let $U \subset \mathbb{C}^n$ be an open ball around 0 and let $X_0 \cup X_1 \subset U$ be as above with $0 \in X_1$. Then either*

    (1) $X_0 \cup X_1 = U$

or (2) *there is a complex line $l \subset \mathbb{C}^n$ through 0 and a smaller ball $U_1 \subset U$ around 0 such that $X_1 \cap l \cap U_1 = \{0\}$ and $X_0 \cap l \cap U_1$ is a countable set of points whose only limit point is 0. In particular, $(X_0 \cup X_1) \cap l \cap U_1$ is compact.*

Proof. If there is a point $P \in U - X_0 \cup X_1$, let l be the line joining O and P. Then $X_1 \cap l$ is an analytic subset of the disc $U \cap l$ containing O but not P. Thus it is a countable discrete set of points and O has a neighborhood U_1 for which $X_1 \cap l \cap U_1$ is only the point O. Similarly, $X_0 \cap l$ is an analytic subset of $U \cap l - X_1 \cap l$, hence is a countable set of points accumulating only at the discrete set of points $X_1 \cap l$.   QED

So assuming $X_0 \cup X_1 \subsetneq U$, choose l as in the lemma and let $p: \mathbb{C}^n \to \mathbb{C}^{n-1}$ be the projection such that $l = p^{-1}(0)$. Since $(X_0 \cup X_1) \cap l \cap U_1$ is compact, by lemma (3.11) there are neighborhoods $U_2 \subset U_1$ and $V \subset \mathbb{C}^{n-1}$ of O such that $p(U_2) \subset V$ and res $p: (X_0 \cup X_1) \cap U_2 \to V$ is proper. Applying (4.11), we conclude that

    i) $p(X_1 \cap U_2)$ is an analytic subset Y_1 of V and res $p: X_1 \cap U_2 \to Y_1$ is finite-to-one,

ii) $p(X_0 \cap U_2 - p^{-1}Y_1)$ is an analytic subset Y_0 of $V - Y_1$ and res $p:(X_0 \cap U_2 - p^{-1}Y_1) \to Y_0$ is finite-to-one,

hence

iii) $p((X_0 \cup X_1) \cap U_2) = Y_0 \cup Y_1$ which is the same kind of subset of V as $X_0 \cup X_1$ was of U.

Moreover

iv) all fibres $p^{-1}(y)$ are countable.

It follows that we can continue by induction and keep on projecting until the image contains a neighborhood of 0. Summarizing, this means that given $X_0 \cup X_1 \subset U \subset \mathbb{C}^n$, where X_1 is closed analytic in U, X_0 is closed analytic in $U - X_1$, and $0 \in X_1$, we can find a projection $p:\mathbb{C}^n \to \mathbb{C}^m$ and open sets $U_1 \subset U, V \subset \mathbb{C}^m$ with $0 \in U_1, p(U_1) \subset V$ such that

i) res $p:(X_0 \cup X_1) \cap U_1 \to V$ is proper and surjective with countable fibres,

ii) $p(X_1 \cap U_1) = Y_1$ is an analytic subset of V and res $p:X_1 \cap U_1 \to Y_1$ is finite-to-one,

iii) res $p:X_0 \cap U_1 - p^{-1}(Y_1) \to V - Y_1$ is finite-to-one.

Returning to the situation and notation of (4.7), we want to apply this to $X_0 = X^{(r)}$ an r-dimensional manifold and $X_1 = X'$ an analytic and *-analytic set of *lower* dimension. In this case, it is pretty clear that the m in question where projection stops will then be r: in fact, for $m > r$, the image of any real-differentiable map from $X^{(r)} \cup X'$ to $\mathbb{C}^m$ could not contain a neighborhood of 0 (because $X^{(r)} \cup X'$ is union of s-dimensional complex manifolds, $s \le r$, hence a union of $2s$-dimensional real manifolds, and the image of k-manifold in an l-manifold under a differentiable map has measure 0 if $k < l$). Now when we are projecting to $\mathbb{C}^r$, at this point $p(X')$ is still an analytic subset of $\mathbb{C}^r$ not containing a neighborhood of 0, hence defined by at least one non-zero equation near 0. But if $p(X^{(r)} \cup X')$ did not contain a neighborhood of 0, we could project one step further and get a countable-to-one differentiable map from $X^{(r)}$ to $\mathbb{C}^{r-1}$: this is absurd (for instance, by Sard's lemma). Thus $m = r$ in our case:

(4.13)

$$
\begin{array}{ccccc}
X^{(r)} - p^{-1}(Y_1) & \subset & X & \supset & X' \\
\downarrow & & \downarrow p & & \downarrow \\
V - Y_1 & \subset & V & \supset & p(X') = Y_1 \\
\text{open, dense} & & \text{neigh. of} & & \text{non-trivial} \\
\text{subset of } V & & 0 \text{ in } \mathbb{C}^r & & \text{analytic subset}
\end{array}
$$

Note that $V - Y_1$ is open dense in V and $X^{(r)} - p^{-1}(Y_1)$ is open dense in $X^{(r)}$. (Both of these are because the complement of non-trivial equation $f = 0$ in a complex manifold is open dense—e.g., join a point x_1 where $f(x_1) \ne 0$ to a point x_2 where $f(x_2) = 0$ by a complex line l and note that $l \cap (f = 0)$ is a discrete set of points.) Also we may assume if we like that V is a ball, in which case it is easy to see that $V - Y_1$ is connected. Next,

$$q = \text{res } p:X^{(r)} - p^{-1}(Y_1) \longrightarrow V - Y_1$$

is a proper, finite-to-one map of the r-dimensional complex manifold to another.

At each point $x \in X^{(r)} - p^{-1}(Y_1)$, one can form the Jacobian of this map

$$J = \frac{\partial(X_1,\dots,X_r)}{\partial(t_1,\dots,t_r)}$$

via coordinates $X_1,\dots,X_r$ on V and local coordinates $t_1,\dots,t_r$ on $X^{(r)}$. Then $J(x) \neq 0$ if and only if q is a local analytic isomorphism near x, i.e., locally has an inverse. If we change coordinates, J changes only by a unit, so that analytic subset B_1 of $X^{(r)} - p^{-1}(Y_1)$ defined by $J = 0$ is independent of choice of coordinates and is a closed analytic subset of $X^{(r)} - p^{-1}(Y_1)$. Therefore by (4.11), $B = q(B_1)$ is an analytic subset of $V - Y_1$. Note that

$$x \notin B \Longleftrightarrow q \text{ is unramified at all points over } x,$$

i.e., B is the branch locus of q. By Sard's lemma, q must be smooth over almost all points of $V - Y_1$, hence B is a non-trivial analytic subset of $V - Y_1$. Thus $V - B - Y_1$ is dense in V and $X^{(r)} - p^{-1}(B \cup Y_1)$ is dense in $X^{(r)}$. Also $V - B - Y_1$ is still connected. We have now constructed the local model of $X^{(r)}$ that we promised—we have a finite unramified covering r:

$$
\begin{array}{ccc}
r = \operatorname{res} q : X^{(r)} - p^{-1}(Y_1 \cup B) & \longrightarrow & V - Y_1 - B \\
\cap & & \cap \\
\mathbb{C}^n & \xrightarrow{\quad p \quad} & \mathbb{C}^r
\end{array}
$$

and $\overline{X^{(r)}}$ is just its closure in $\mathbb{C}^n$. The problem now is—what can we say about this closure? We shall analyze it at this point by constructing explicitly some quite simple analytic functions on $p^{-1}(V)$ that vanish on $\overline{X^{(r)}}$:

(1) let d be the number of sheets in the covering r

(2) choose a linear function l on $\mathbb{C}^n$

(3) for all $i, 1 \leq i \leq d$, and all $y \in V - Y_1 - B$, let $a_i(y)$ be the ith elementary symmetric function of the numbers $l(x_1),\dots l(x_d)$ where $\{x_1,\dots,x_d\} = r^{-1}(y)$. Using local inverses for r near y, it follows that as y moves, the x_i vary analytically, hence so do $l(x_i)$, hence so does $a_i(y)$. Thus a_i is an analytic function on $V - Y_1 - B$.

(4) The functions a_i extend to analytic functions on V. In fact, by (4.10), it suffices to show that the a_i are bounded on $K \cap (V - Y_1 - B)$ for every compact set $K \subset V$. But by the properness of res $p : (X^{(r)} \cup X') \to V, (X^{(r)} \cup X') \cap p^{-1}K$ is compact, hence $l(x)$ is bounded on $X^{(r)} \cap p^{-1}K$, hence the a_i are bounded on $K \cap (V - Y_1 - B)$.

(5) Let F_l be the analytic function on $p^{-1}(V)$

$$F_l(x) = l(x)^d - a_1(p(x)) \cdot l(x)^{d-1} + \dots + (-1)^d \cdot a_d(p(x)).$$

It is clear that $F_l \equiv 0$ on $X^{(r)} - p^{-1}(Y_1 \cup B)$, hence $F_l \equiv 0$ on $\overline{X^{(r)}}$.

Now everything falls out. We can complete the proof of (4.7), hence of (4.5), by checking that $\overline{X^{(r)}}$ is the set of points in $p^{-1}(V)$ where all the functions F_l are zero. In fact, we will see that a finite number of F_l's suffice. Let $x \in p^{-1}V - \overline{X^{(r)}}$, let $y = f(x)$ and let $y = \lim y_k, y_k \in V - B - Y_1$. Then $r^{-1}(y_k) = \{x_k^{(1)},\dots,x_k^{(d)}\}$ and because res $p : \overline{X^{(r)}} \to V$ is proper, we may pass to a subsequence such that for all

$j = 1, \ldots, d, x_k^{(j)}$ has a limit $x^{(j)}$ as $k \to \infty$. Then $x^{(j)} \in \overline{X^{(r)}}$ so $x \neq x^{(j)}$. Now choose*
l such that $\bar{l}(x) \neq l(x^{(j)})$ for any j. Then note that by definition $l(x_k^{(1)}), \ldots, l(x_k^{(d)})$
are the complete set of roots of the polynomial

$$t^d - a_1(y_k)t^{d-1} + \ldots + (-1)^d \cdot a_d(y_k).$$

Therefore $l(x^{(1)}), \ldots, l(x^{(d)})$ are the only roots of

$$t^d - a_1(y)t^{d-1} + \ldots + (-1)^d \cdot a_d(y).$$

Therefore

$$F_l(x) = l(x)^d - a_1(y) \cdot l(x)^{d-1} + \ldots + (-1)^d \cdot a_d(y) \neq 0.$$

This same proof shows also that

(*) $\overline{X^{(r)}} \cap p^{-1}(y) = \{x^{(1)}, \ldots, x^{(d)}\}$

since it actually shows that if $x \neq x^{(j)}$, any j, and $p(x) = y$, then $F_l(x) \neq 0$, hence
$x \notin \overline{X^{(r)}}$. Thus

$$\#\left[\overline{X^{(r)}} \cap p^{-1}(y)\right] \leq d, \quad \text{all } y \in V,$$

and thus we also see that *all* fibres of

$$\text{res } p : X^{(r)} \cup X' \longrightarrow V$$

were actually finite all along: a sticky point that we couldn't prove earlier. It
also shows that

$$\text{res } p : \overline{X^{(r)}} \longrightarrow V$$

is an *open* map, just as in Theorem (3.10). In fact, suppose $x \in \overline{X^{(r)}}$ and U is a neighborhood of $x \in \overline{X^{(r)}}$ such that $p(U)$ is not a neighborhood of $y = p(x)$: then we
could choose a sequence $y_k \in V - p(U)$ with $\lim y_k = y$, and proceeding as above
we would get $x_k^{(j)} \notin U$, hence their limits $x^{(j)}$ would be distinct from x, which would
contradict (*).

§4B. Applications to Uniqueness of Algebraic Structure and Connectedness

In addition to its intrinsic beauty, Chow's theorem (4.6) has several important
applications. The first of these is:

(4.14) **Corollary.** *Let X and Y be smooth projective varieties. Then every holomorphic map*

$$f : X \longrightarrow Y$$

is a regular correspondence from X to Y and conversely.

Proof. Given f, let $\Gamma_f \subset X \times Y$ be its graph. By Chow's theorem, Γ_f is a

---

*In fact, say $\{l_\alpha\}_{\alpha \in I}$ are any $(n - r - 1)d + 1$ linear functions "in general position" with respect to the
$n - r$-dimensional linear subspace $p^{-1}o$, i.e., any subset of $n - r$ of them are independent on $p^{-1}o$.
Then one of these l_α is OK. If not, for all α, there would be $j(\alpha)$ such that $l_\alpha(x) = l_\alpha(x^{(j(\alpha))})$. Some j_0 would
have to occur $n - r$ times as $j(\alpha)$, so $l_\alpha(x) = l_\alpha(x^{(j_0)})$ for $n - r$ α's. Since these l_α are independent in the
coset $p^{-1}(p(x))$ of $p^{-1}o$ containing x and $x^{(j_0)}$, $x = x^{(j_0)}$ — a contradiction.

closed algebraic subset of $X \times Y$. Since $p_1 : \Gamma_f \to X$ is bijective and closed in Zariski topology, Γ_f is irreducible in the Zariski topology, hence Γ_f is a rational correspondence from X to Y. By Zariski's Main Theorem, Γ_f is regular. The converse is obvious.   QED

(4.15) **Corollary.** *Let X and Y be smooth projective varieties. Then X and Y are isomorphic as complex manifolds if and only if there exists a biregular correspondence $Z : X \to Y$.*

This result is often paraphrased more succinctly by saying: every compact complex manifold has at most one "algebraic structure". It is also true, but not at all easy to prove, that every compact 1-*dimensional* complex manifold has an algebraic structure, i.e., is analytically isomorphic to a complex projective variety (cf. Weyl [1]); but that for all dimensions bigger than 1, there are non-algebraic compact complex manifolds and even ones with no non-constant meromorphic functions on them at all.

Chow's theorem also gives us the following connectedness result:

(4.16) **Corollary.** *Let $X \subset \mathbb{P}^n$ be an r-dimensional projective variety and let $Y \subsetneqq X$ be a closed algebraic set. Then $X - Y$ is connected in the classical topology.*

Proof. If $X - Y = Z_1 \cup Z_2, Z_i$ open and closed in $X - Y$, then let $S = \operatorname{Sing} X$ and apply Chow's theorem to $X = Z_1 \cup Y \cup S$. This set can be stratified by letting:

$$X^{(r)} = Z_1 - Z_1 \cap (S \cup Y)$$
$$X^{(i)} = \text{suitable stratification of the algebraic set } S \cup Y \text{ for } 0 \leq i \leq r - 1.$$

It follows that $Z_1 \cup S \cup Y$ is algebraic. Similarly $Z_2 \cup S \cup Y$ is algebraic, hence

$$X = (Z_1 \cup S \cup Y) \cup (Z_2 \cup S \cup Y)$$

hence X is not irreducible.   QED

We can get an even stronger connectedness assertion about X as follows:

(4.17) **Proposition.** *Let $X \subset \mathbb{P}^n$ be an r-dimensional projective variety and let $M^{n-r-1} \subset \mathbb{P}^n$ be a linear space disjoint from X. Let $p : X \to \mathbb{P}^r$ be projection from M and let*

$$B = \{x \in \mathbb{P}^r \mid p \text{ not smooth over } x\}$$

so that

$$(X - p^{-1}B) \longrightarrow \mathbb{P}^r - B$$

is a finite-sheeted connected covering space. Let $l \subset \mathbb{P}^r$ be any line which meets B transversely, (i.e., l meets B only at smooth points $x \in B$, on $(r-1)$-dimensional components of B, with $T_{x,l}, T_{x,B}$ transversal). Then

$$p^{-1}(l - l \cap B) \longrightarrow l - l \cap B$$

is also a connected covering space, hence $p^{-1}(l)$ is an irreducible curve.

Proof. Take any point $x \in l - l \cap B$ and look at $p_x : \mathbb{P}^r - \{x\} \to \mathbb{P}^{r-1}$. p_x restricts to a projection

$$\text{res } p_x : B \longrightarrow \mathbb{P}^{r-1},$$

and let $B_0 \subset \mathbb{P}^{r-1}$ be the set of points where res p_x is not smooth. Then

$$B - p_x^{-1}(B_0) \longrightarrow \mathbb{P}^{r-1} - B_0$$

is a finite-sheeted covering space of $(r-1)$-dimensional manifolds, say of degree e. For all $y \in \mathbb{P}^{r-1}$, let l_y be the line $p_x^{-1}(y) \cup \{x\}$. Then, in fact, $\mathbb{P}^{r-1} - B_0 = \{y \in \mathbb{P}^{r-1} | l_y$ meets B transversely$\}$. As y varies over $\mathbb{P}^{r-1} - B_0$, $l_y \cap B$ is a continuously varying set of e points of l_y, hence the open manifolds $l_y - l_y \cap B$ are all diffeomorphic to each other or to a 2-sphere minus e points. Now consider the continuous family of coverings

$$p^{-1}(l_y - l_y \cap B) \longrightarrow l_y - l_y \cap B.$$

Suppose $l = l_{y_0}$ and $p^{-1}(l - l \cap B)$ is disconnected. Then as y varies over $\mathbb{P}^{r-1} - B_0$, all of these coverings are disconnected. Choose a point $z \in p^{-1}(x)$ and let

$$p^{-1}(l_y - l_y \cap B)_1 = \binom{\text{the connected component of}}{p^{-1}(l_y - l_y \cap B) \text{ through } z},$$

$$p^{-1}(l_y - l_y \cap B)_2 = (\text{the other components}).$$

Let

$$B^* = B \cup \{x\} \cup p_x^{-1}(B_0).$$

Then it follows that $X - p^{-1}(B^*)$ breaks up into 2 open and closed pieces:

$$\bigcup_{y \in \mathbb{P}^{r-1} - B_0} p^{-1}(l_y - l_y \cap B)_1 - p^{-1}(x)$$

and

$$\bigcup_{y \in \mathbb{P}^{r-1} - B_0} p^{-1}(l_y - l_y \cap B)_2 - p^{-1}(x)$$

and this contradicts (4.16).   QED

(4.18) **Corollary.** *Given $X^r \subset \mathbb{P}^n$, there is a linear subspace $L^{n-r+1} \subset \mathbb{P}^n$ such that $X \cap L$ is an irreducible curve and X, L meet transversely (i.e., $\exists z \in X \cap L, z$ smooth on X and $T_{z,X}, T_{z,L}$ transverse linear subspaces of $T_{z,\mathbb{P}^n}$).*

Proof. In the notation of (4.17), take an l transversal to B (these exist because $B_0 \subsetneqq \mathbb{P}^{r-1}$ and l_y is transversal to B if $y \notin B_0$). Let $L = p_M^{-1}(l) \cup M$ be the corresponding linear subspace of $\mathbb{P}^n$. Then dim $L = n - r + 1$, $L \cap X = p^{-1}(l)$ and if $z \in p^{-1}(l - l \cap B)$ the transversality at z follows because via $dp_M : T_{z,\mathbb{P}^n} \to T_{z,\mathbb{P}^r}$

$$dp_M : T_{z,X} \xrightarrow{\approx} T_{z,\mathbb{P}^r} \quad \text{is an isomorphism,}$$

and

$$T_{z,L} = (dp_M)^{-1}(T_{z,l}). \quad \text{QED}$$

Chapter 5. Degree of a Projective Variety

§ 5A. Definition of $\deg(X)$, $\mathrm{mult}_x X$, of the Blow-up $B_x(X)$, Effect of a Projection, Examples

Let $X^r \subset \mathbb{P}^n$ be a variety of dimension r. The degree of X is to be the number of points in which almost all linear subspaces $L^{n-r} \subset \mathbb{P}^n$ meet X. (Note that here and in what follows, we use the notation Y^s to denote an s-dimensional variety). More precisely, we make the definition via a theorem:

(5.1) **Theorem.** *For all subvarieties $X^r \subset \mathbb{P}^n$, there exists an integer $d \geq 1$ such that: if $L^{n-r} \subset \mathbb{P}^n$ is a linear subspace satisfying:*
   a) $L \cap X = finite\ set\ \{x_1, \ldots, x_k\}$,
   b) $\forall\ i,\ x_i$ *is smooth on X and the 2 subspaces*

$$T_{x_i, X}, T_{x_i, L}\ of\ T_{x_i, \mathbb{P}^n}\ meet\ only\ in\ 0,$$

then $k = d$.

Before proving the theorem, note that such L's exist: in fact let M^{n-r-1} be a linear space of dimension one less than L disjoint from X, and let projection from M define $\pi = \mathrm{res}\ p_M$:

$$\mathbb{P}^n - M \xrightarrow{\ p_M\ } \mathbb{P}^r$$
$$\cup$$
$$X \xrightarrow{\qquad \pi \qquad}$$

For almost all $y \in \mathbb{P}^r$, π is smooth over y. If π is smooth over y, let $L = M \cup p_M^{-1}(y)$. L is a linear space of dimension $n - r$ such that $L \cap X = \pi^{-1}(y)$. If $x \in \pi^{-1}(y)$, look at the differential

$$dp_M : T_{x, \mathbb{P}^n} \longrightarrow T_{y, \mathbb{P}^r}\ .$$

Then $\ker(dp_M) = T_{x, L}$ and because $d\pi$ is smooth, dp_M restricts to an isomorphism of $T_{x, X}$ onto $T_{y, \mathbb{P}^r}$. Therefore $T_{x, L} \cap T_{x, X} = (0)$. We will summarize properties (a) and (b) by saying that X and L *meet transversely.*

Proof of (5.1). Introduce new variables $\xi_j^{(i)}$, $1 \leq i \leq r$, $0 \leq j \leq n$. Let

$$L_\xi = V\left(\sum_{j=0}^n \xi_j^{(1)} X_j, \ldots, \sum_{j=0}^n \xi_j^{(r)} X_j \right)$$

be the linear space defined by r equations cooked up from the ξ's. Among the L_ξ's, we get of course all linear subspaces of $\mathbb{P}^n$ of dimension $n - r$, as well as those of higher dimensions. We want to look at the intersections $X \cap L_\xi$, as ξ varies, by fitting all the sets $X \cap L_\xi$ together into a variety $\mathfrak{X}$ "lying over" the affine ξ-space, with $X \cap L_\xi$ as its fibres over various points. To be precise, let $\mathbb{C}_\xi^{r(n+1)}$ be affine space with coordinates $\xi_j^{(i)}$ and let

$$\mathfrak{X} = \left\{ (x,\xi) \,\middle|\, x \in X, \sum_{j=0}^{n} \xi_j^{(i)} x_j = 0, 1 \le i \le r \right\} \subset \mathbb{P}^n \times \mathbb{C}_\xi^{r(n+1)}$$

Let $p : \mathfrak{X} \to \mathbb{C}_\xi^{r(n+1)}$ be the projection. Then indeed $p^{-1}(\xi) = X \cap L_\xi$. By (3.30), this is always non-empty, hence p is surjective. Now $\mathfrak{X}$ is a closed algebraic subset of $X \times \mathbb{C}_\xi^{r(n+1)}$ defined locally by r equations: therefore by (3.14), all components of $\mathfrak{X}$ have dimension at least $r(n+1)$. The key point is:

(5.2) **Lemma.** *For all $\xi \in \mathbb{C}^{r(n+1)}$,*

$$\begin{bmatrix} X \text{ meets } L_\xi \text{ transversely in a} \\ \text{finite set of points} \end{bmatrix} \Longrightarrow \begin{bmatrix} \text{each point } x \in p^{-1}(\xi) \text{ is on a unique} \\ \text{component of } \mathfrak{X} \text{ which is smooth over} \\ \mathbb{C}^{r(n+1)} \text{ at } x \end{bmatrix}$$

Proof. Say X meets L_ξ transversely and $x \in X \cap L_\xi$. Assume $X_0 \ne 0$ at x. Since x is smooth on X, let $f_1, \dots, f_{n-r}$ be polynomials with independent differentials at x defining X near x. Let $L_i(\xi, X) = \sum \xi_j^{(i)}(X_j/X_0)$. The differential dL_i is a linear form on $T_{(\xi,x),\mathbb{C}^{r(n+1)} \times \mathbb{P}^n} = T_{\xi,\mathbb{C}^{r(n+1)}} \times T_{x,\mathbb{P}^n}$. Now as linear forms on the subspace $(0) \times T_{x,\mathbb{P}^n}$, $dL_1 = \dots = dL_r = 0$ defines the tangent space T_{x,L_ξ} to L_ξ. Since X and L_ξ meet transversely, it follows that $df_1, \dots, df_{n-r}, dL_1, \dots, dL_r$ are independent linear forms on $(0) \times T_{x,\mathbb{P}^n}$, hence a fortiori are independent linear forms on $T_{(\xi,x),\mathbb{C}^{r(n+1)} \times \mathbb{P}^n}$. Therefore $f_1, \dots, f_{n-r}, L_1, \dots, L_r$ define a variety Z through (ξ,x) of dimension $(r(n+1)+n)-n = r(n+1)$, with a smooth point at (ξ,x). Since $Z \supset \mathfrak{X}$ near (ξ,x) and all components of $\mathfrak{X}$ have dimension at least $r(n+1)$, it follows that $Z = \mathfrak{X}$ near (ξ,x), hence $\mathfrak{X}$ has a unique component through (ξ,x) and $T_{(\xi,x),Z}$ is the set of zeroes of df_i and dL_j. But the kernel of $dp : T_{(\xi,x),Z} \to T_{\xi,\mathbb{C}^{r(n+1)}}$ is the intersection $(0) \times T_{x,\mathbb{P}^n} \cap T_{(\xi,x),Z}$ and we just saw that the df_i and dL_j have no common zeroes except $(0,0)$ in $(0) \times T_{x,\mathbb{P}^n}$. Therefore p is smooth at (x,ξ). QED

The Theorem now follows immediately from the Specialization principle (3.26). In fact (3.26) also implies:

(5.3) **Corollary of proof.** *Let $\mathscr{Z}_0(\mathbb{P}^n) = $ free abelian group generated by the points of $\mathbb{P}^n$. Then if x is an isolated point of $X \cap L_\xi$, define*

$$i(x; X \cap L) = \text{mult}_{(x,\xi)} \, p$$

and

$$X \cdot L = \sum_{x \in X \cap L} i(x; X \cap L) \cdot x \in \mathscr{Z}_0(\mathbb{P}^n).$$

Equivalently, going back to the definition of mult, this means that for every sufficiently small classical neighborhood V of $x \in \mathbb{P}^n$, there is a neighborhood U of $\xi \in \mathbb{C}_\xi^{r(n+1)}$ such that if $\xi' \in U$ and $L_{\xi'} \cap X$ is transversal, then

$$i(x, X \cap L) = \#(X \cap L_{\xi'} \cap V).$$

Then for all linear spaces L of dimension $n - r$ such that $X \cap L$ is finite,

$$\deg(X \cdot L) = \deg X,$$

where for any element $\sum n_i P_i \in \mathscr{Z}_0(\mathbb{P}^n)$, $\deg(\sum n_i P_i) = \sum n_i$.

A more direct way of defining $i(x; X \cap L)$ by means of mult is perhaps to choose $M^{n-r-1} \subset L$ such that $x \notin M$ and let $p_M : \mathbb{P}^n - M \to \mathbb{P}^r$ be the projection. Then

(5.4) $$i(x; X \cap L) = \text{mult}_x(\text{res}_X \, p_M).$$

The equality here is immediate because by definition, $\text{mult}_x(\text{res}_X \, p_M) = \#(X \cap p_M^{-1}(y) \cap V)$ for y near $p_M(x)$, and $p_M^{-1}(y) = L_{\xi'}$ for ξ' near ξ.

The simplest example of the theorem is the case of a hypersurface $X = V(f)$, f an irreducible homogeneous polynomial of degree d. If $x, y \in \mathbb{P}^n$, the line $\overline{xy}$ joining x and y is the set of points with homogeneous coordinates $sx + ty$ (vector notation). Therefore the intersections $\overline{xy} \cap X$ are in $1 - 1$ correspondence with the roots of $g(s,t) = f(sx + ty)$, a homogeneous polynomial in s and t of degree d. If $\overline{xy}$ meets X transversely, it is easy to check that the roots of g are all simple; hence there are d of them, and d is the degree of X. Moreover, for any line $\overline{xy}$ such that $\overline{xy} \not\subset X$, then $g \not\equiv 0$ and d is still the total number of roots if they are counted with their multiplicities, i.e.,

if $g(s,t) = \prod (\alpha_i s + \beta_i t)^{r_i}$,   all α_i/β_i distinct

and $P_i = \alpha_i x + \beta_i y$,

then $X \cap \overline{xy} = \{ \ldots, P_i, \ldots \}$

and $d = \sum r_i$.

Now if x and y are moved slightly to points x_α, y_α so that $\overline{x_\alpha y_\alpha}$ meets X transversely, then $g_\alpha(s,t) = f_\alpha(sx_\alpha + ty_\alpha)$ has d distinct roots of which r_i approach $(\alpha_i : \beta_i)$ as $x_\alpha \to x, y_\alpha \to y$. Thus r_i points in $X \cap \overline{x_\alpha y_\alpha}$ approach P_i and hence

$$r_i = i(P_i ; X \cap \overline{xy}).$$

and

$$X \cdot \overline{xy} = \sum_i r_i P_i.$$

In proving theorems about degree, a useful technique is to analyze intersections with a "sufficiently general" linear space L. The question often is how we know that such a space exists. To check this in any particular case, it will suffice to check that each requirement on L is satisfied by all L_ξ for ξ in a non-empty Zariski open subset of $\mathbb{C}^{r(n+1)}$. Since a finite intersection of non-empty Zariski opens is still Zariski open *and non-empty*, plenty of good L's will always exist. For instance, the set of all ξ such that X meets L_ξ transversely is easily checked to be Zariski open as well as non-empty. Here is an example of this technique:

(5.5) **Proposition.** *Let* $X^r \subset \mathbb{P}^n - \{x\}$ *and let* $p_x : \mathbb{P}^n - \{x\} \to \mathbb{P}^{n-1}$ *be the projection. Let* $Y = p_x(X)$, *let* $\pi = \text{res} \, p_x : X \to Y$. *Then*

$$\deg X = \deg \pi \cdot \deg Y.$$

Proof. Let $S = \left\{ y \in Y \, \middle| \, \begin{array}{l} y \text{ is singular on } Y \text{ or} \\ \pi \text{ is not smooth over } y \end{array} \right\}.$

Let $L^{n-r-1} \subset \mathbb{P}^{n-1}$ be a linear space such that:
  a) $L \cap S = \phi$
  b) L meets Y transversely.
Let $M = \{x\} \cup p_x^{-1}(L) : M$ is an $n-r$-dimensional linear subspace of $\mathbb{P}^n$ such that $M \cap X = \pi^{-1}(L \cap Y)$. Because of (a), π is smooth at all points of $M \cap X$. It is easy to check that because of (a) and (b), M meets X transversely. Then

$$\begin{aligned}
\deg X &= \#(M \cap X) \\
&= \deg \pi \cdot \#(L \cap Y) \\
&= \deg \pi \cdot \deg Y. \quad \text{QED}
\end{aligned}$$

Applying (5.5) inductively, we are led to:

(5.6) **Corollary.** *Let L^{n-r-1} be disjoint from X^r and let $\pi = \mathrm{res}(p_L)$ be the projection from X to $\mathbb{P}^r$. Then*

$$\deg X = \deg \pi.$$

On the other hand, if $x \in X^r \subset \mathbb{P}^n$ and we project X^r from x some quite extraordinary things happen. To work this out, we must make a digression on the infinitesimal structure of X near x. First of all, recall from Chapter 2 that the map $p_x : \mathbb{P}^n - \{x\} \to \mathbb{P}^{n-1}$ is part of a rational map:

$$\Gamma \subset \mathbb{P}^n \times \mathbb{P}^{n-1}$$

where

$$\Gamma \cap [(\mathbb{P}^n - \{x\}) \times \mathbb{P}^{n-1}] = \text{graph of } p_x$$
$$\Gamma \cap [\{x\} \times \mathbb{P}^{n-1}] = \underbrace{\{x\} \times \mathbb{P}^{n-1}}$$

call this E.

Note that the points of E are in $1-1$ correspondence with the points of $\mathbb{P}^{n-1}$, hence via p_x^{-1}, with the lines $l \subset \mathbb{P}^n$ through x. To describe Γ as a topological space, in the classical topology, we may say:
  a) take $\mathbb{P}^n - \{x\}$ as an open part,
  b) take the $\mathbb{P}^{n-1}$ parametrizing lines through x as a closed part,
  c) topologize the union by saying that $y_i \in \mathbb{P}^n - \{x\}, i = 1, 2, \dots$ converge to a line l in the $\mathbb{P}^{n-1}$ piece if $y_i \to x$ as $i \to \infty$ and the lines $\overline{y_i x}$ converge to l, [i.e., if $X_1, \dots, X_n$ are affine coordinates in $\mathbb{P}^n$ with x as origin, then this means $\exists \lambda_i \in \mathbb{C}$ such that $\lambda_i \to \infty$ but $(\lambda_i X_1(y_i), \dots, \lambda_i X_n(y_n)) \to (\alpha_1, \dots, \alpha_n) \in \mathbb{C}^n$ and $l = $ locus of points $(\alpha_1 t, \dots, \alpha_n t)$.]

It is intuitively clear to the reader, I hope, why Γ is called the variety obtained by "*blowing up x*" on $\mathbb{P}^n$. Another important property of Γ is that at every point $z \in E$, E is defined as a subset of Γ as the zeroes of one function in some neighborhood of z. In fact, let $X_0, \dots, X_n$, resp. $Y_1, \dots, Y_n$ be homogeneous coordinates on $\mathbb{P}^n$, resp. $\mathbb{P}^{n-1}$, such that $x = (1, 0, \dots, 0)$ and Γ is the closure of the graph of the map $Y_i = X_i, 1 \leq i \leq n$. If $z = (x, y) \in E$, and $Y_i(y) \neq 0$, then

$$\begin{aligned}
V(X_i) \cap \Gamma &= \{(x', y') \in \Gamma \mid X_i(x') = 0\} \\
&= \{(x', y') \in \Gamma \mid x' = x \text{ or } Y_i(y') = 0\} \\
&= E \cup [\Gamma \cap V(Y_i)].
\end{aligned}$$

Since $z \notin \Gamma \cap V(Y_i)$, E is defined near z by the single equation $X_i / X_0 = 0$.

Now let's take up again an r-dimensional variety $X^r \subset \mathbb{P}^n$ containing x.

(5.7) **Definition.** $B_x(X) = Zariski$ $closure$ in Γ of $X - \{x\}$ $(more$ $precisely,$ of
$$\{(x', p_x x') \mid x' \in X - \{x\}\})$$

$$p_x(X) = projection\ p_2(B_x(X))\ in\ \mathbb{P}^{n-1}$$
$$= Zariski\text{-}closure\ of\ p_x(X - \{x\}).$$

$$E_{x,X} = E \cap B_x(X).$$

$$E^*_{x,X} = \{y \in \mathbb{P}^n \mid y = x\ or\ (x, p_x y) \in E_{x,X}\}$$
$$= p_x^{-1}(p_2(E_{x,X})) \cup \{x\}.$$

Just as with $\Gamma, B_x(X)$ as a topological space in the classical topology can be constructed as follows:

a) take $X - \{x\}$ as an open piece,

b) take the set $E_{x,X}$ of lines l through x which are limits of secants $\overline{xy_i}$, $y_i \in X - \{x\}$ and $y_i \to x$ as $i \to \infty$ (topologized as a subset of the $\mathbb{P}^{n-1}$ of all lines through x)

c) topologize the result to that now $y_i \to l$ as $i \to \infty$

$B_x(X)$ is called the *variety obtained by blowing up* x *on* X; $E_{x,X}$ is called the *projectivized tangent cone at* x; and $E^*_{x,X}$ is called the *tangent cone at* x. $B_x(X)$ is an r-dimensional variety and

$$\text{res } p_1 : B_x(X) \longrightarrow X$$

is a birational map. Moreover, since E is defined in Γ locally by one equation, so is $E_{x,X}$ in $B_x(X)$. Therefore, by (3.14), $E_{x,X}$ is a union of r-1-dimensional components.

Finally, near x, I say that X approximates its tangent cone very closely. In fact if we introduce a positive definite quadratic form in the real vector space $T_{x,\mathbb{P}^n}$, we can define neighborhoods of the tangent cone:

$$E^*_{x,X}(\varepsilon) = \left\{ y \in \mathbb{P}^n - \{x\} \,\middle|\, \exists z \in E^*_{x,X} \text{ such that the angle } \theta \text{ in } y \text{ satisfies } |\theta| < \varepsilon \right\} \cup \{x\}$$

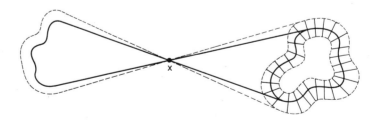

Then using the fact that $E^*_{x,X}$ contains the limits of all secants $\overline{xy}$, it follows easily that for all ε, there is a neighborhood U of $x \in \mathbb{P}^n$ such that

(5.8) $X \cap U \subset E^*_{x,X}(\varepsilon).$

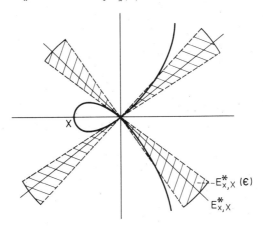

We now make the following definition:

(5.9) **Definition.** $\text{mult}_x(X^r) = \min\limits_{\left[\substack{L^{n-r}\text{ linear space}\\ s.t.\{x\}=\text{comp.}\\ \text{of }X\cap L}\right]} i(x;X\cap L).$

We can illustrate these definitions again with the case of a hypersurface $X = V(F)$. If $x \in X$ and $X_1,\ldots,X_n$ are affine coordinates with $x = $ origin, let

$$f(X_1,\ldots,X_n) = F/l^d, \qquad \begin{aligned} d &= \text{degree } F\\ l &= \text{linear equation of}\\ &\quad\text{hyperplane at } \infty\end{aligned}$$

be the affine equation of X. If we write

$$f = f_r + f_{r+1} + \ldots + f_d$$

$$f_i \text{ homogeneous in } X_1,\ldots,X_n \text{ of degree } i, f_r \not\equiv 0$$

then:

(5.10) **Proposition.** i) $E^*_{x,X} = $ *locus of zeroes of the so-called leading term* f_r
ii) $\text{mult}_x X = r$

Proof. We must show that the line $l_a = \{(a_1 t,\ldots,a_n t)\,|\,t \in \mathbb{C}\}$ is a limit of secants $\overline{0,x^{(k)}}$ of X iff $f_r(a_1,\ldots,a_n) = 0$. If $x^{(k)} = (\lambda_k a_1^{(k)},\ldots,\lambda_k a_n^{(k)}) \in X$ where $a_i^{(k)} \to a_i$ and $\lambda_k \to 0$ as $k \to \infty$, then

$$\lambda_k^r f_r(a^{(k)}) + \lambda_k^{r+1} f_{r+1}(a^{(k)}) + \ldots + \lambda_k^d f_d(a^{(k)}) = 0.$$

Dividing by λ_k^r and letting $k \to \infty$, it follows that $f_r(a) = 0$. Conversely if $f_r(a) = 0$, look at $(a_1,\ldots,a_n,0)$ as a solution of

$$f_r(X_1,\ldots,X_n) + Y f_{r+1}(X_1,\ldots,X_n) + \ldots + Y^{d-r} f_d(X_1,\ldots,X_n) = 0.$$

By (2.33) this solution is a limit of solutions $(a_1^{(k)},\ldots,a_n^{(k)},\lambda_k)$ with $\lambda_k \neq 0$. If $x^{(k)} = (\lambda_k a_1^{(k)},\ldots,\lambda_k a_n^{(k)})$, then $x^{(k)} \in X$ and $\overline{0,x^{(k)}} \to l_a$ as $k \to \infty$. As for $\text{mult}_x X$, it is easy to check that if $f_r(a) \neq 0$, then $l_a \cap X$ meets X r times at x and $d - r$ times elsewhere.

QED

[By the same method, one can show that for any $X \subset \mathbb{C}^N$ if $x = (0,\ldots,0) \in X$ and $X = V(\mathfrak{A})$, then

$$E_{x,X}^* = \begin{cases} \text{locus of common zeroes of the "leading terms"} \\ f_r \text{ of all } f \in \mathfrak{A}. \end{cases} \Big]$$

We can now give a complete description of the effect on degree of projecting from a point of X:

(5.11) **Theorem.** *For all* $x \in X^r \subset \mathbb{P}^n$,

a) $\deg X - \text{mult}_x(X) = \begin{cases} \deg p_x(X) \cdot \deg(\text{res}_X p_x), \text{ if } X \text{ is not} \\ \qquad \text{itself a cone over } x \\ 0 \text{ if } X \text{ is a cone over } x \end{cases}$

b) *for all* L^{n-r} *through* x *such that* $L \cap E_{x,X}^* = \{x\}$,

$$\text{mult}_x(X) = i(x; X \cap L),$$

hence for all M^{n-r-1} *disjoint from* $E_{x,X}^*$, *let* $p_M : \mathbb{P}^n - M \to \mathbb{P}^r$ *be the projection, then*

$$\text{mult}_x(X) = \text{mult}_x(\text{res}_X p_M).$$

Proof. First of all, say X is a cone through x, i.e., $X = E_{x,X}^*$. Then every L^{n-r} through x either contains a whole line of X or meets X only at x. In the second case

$$\deg X = \sum_{y \in X \cap L} i(y, X \cap L) = i(x, X \cap L),$$

hence

$$\deg X = \text{mult}_x(X)$$

too. Now suppose X is not a cone. Then almost all fibres of

$$\pi = \text{res}(p_x) : X \longrightarrow p_x(X)$$

are finite, hence $\dim p_x(X) = r$. Let $B \subset p_x(X)$ be a closed algebraic set outside of which π is smooth and containing $p_2(E_{x,X})$. Then call "good" the linear spaces $M_1^{n-r-1} \subset \mathbb{P}^{n-1}$ such that $M_1 \cap B = \phi$ and M_1 meets $p_x(X)$ transversely. It follows as in (5.5) that if M_1 is good then $L_1 = p_x^{-1}(M_1) \cup \{x\}$ meets X transversely at all points except x and that the number of intersections is $\deg p_x(X) \cdot \deg \pi$. Thus:

(*) $\deg X - i(x, X \cap L_1) = \deg p_x(X) \cdot \deg \pi.$

We obtain in this way a Zariski-open dense set in the set of all linear L^{n-r}'s containing x such that (*) holds. Call these the "good" L^{n-r}'s. But if L is an arbitrary $n-r$-dimensional linear space such that $\{x\}$ is a component of $L \cap X$, then say $L = L_{\xi_0}$ and consider the intersections $L_\xi \cap X$ for ξ near ξ_0. By definition of "mult", there is an open set $U \subset \mathbb{C}_\xi^{r(n+1)}$ containing ξ_0 and a neighborhood $V \subset \mathbb{P}^n$ of x such that

$$\xi \in U, X \cap L_\xi \text{ transversal} \Longrightarrow \#(X \cap L_\xi \cap V) = i(x; L_{\xi_0} \cap X).$$

It follows that

$$\xi \in U, X \cap L_\xi \text{ finite} \Longrightarrow \sum_{y \in L_\xi \cap X \cap V} i(y; L_\xi \cap X) = i(x; L_{\xi_0} \cap X).$$

But there is a $\xi_1 \in U$ such that $x \in L_{\xi_1}$ and L_{ξ_1} is good. Therefore, we find:

$$i(x, L_{\xi_0} \cap X) = \sum_{y \in L_{\xi_1} \cap X \cap V} i(y, L_{\xi_1} \cap X)$$

(**)
$$= i(x, L_{\xi_1} \cap X) + \sum_{\substack{y \neq x \\ y \in L_{\xi_1} \cap X \cap V}} i(y, L_{\xi_1} \cap X).$$

$$\geq i(x, X \cap L_{\xi_1})$$

Therefore min $i(x; X \cap L)$ is taken on for good L's, hence by (*):

$$\text{mult}_x(X) = \deg X - \deg p_x(X) \cdot \deg \pi.$$

But finally if $L_{\xi_0} \cap E^*_{x,X} = \{x\}$, then for some ε

$$L_{\xi_0} \cap E^*_{x,X}(\varepsilon) = \{x\}.$$

Suppose W is a neighborhood of $x \in \mathbb{P}^n$ such that $X \cap W \subset E^*_{x,X}(\varepsilon)$ as in (5.8). As before, approximate ξ_0 by a $\xi_1 \in U$ such that $x \in L_{\xi_1}$ and L_{ξ_1} is good. If ξ_1 is close enough to ξ_0, then:

$$L_{\xi_1} \cap E^*_{x,X}(\varepsilon) = \{x\}$$

$$\text{and} \quad L_{\xi_1} \cap X \cap V \subset W,$$

hence in fact

$$L_{\xi_1} \cap X \cap V \subset L_{\xi_1} \cap X \cap W \subset L_{\xi_1} \cap E^*_{x,X}(\varepsilon) = \{x\}$$

and by (**), $i(x, L_{\xi_0} \cap X) = i(x, L_{\xi_1} \cap X) = \text{mult}_x(X)$. The last assertion follows by (5.4).   QED

(5.12) **Corollary.** *If $X \subset \mathbb{P}^n$ is a projective variety, then $\deg X = 1$ if and only if X is a linear space.*

Proof. Clearly X linear $\Rightarrow \deg X = 1$. Conversely, if $\deg X = 1$, and $x \in X$ then by (5.11), X is a cone over x. This is true for all x: i.e., for all $x, y \in X$, $\overline{xy} \subset X$. This means that X is linear.

(5.13) **Corollary.** *If $X \subset \mathbb{P}^n$ is a projective variety and $X \not\subset$ (any hyperplane), then*

$$\deg X \geq \text{codim } X + 1.$$

Proof. We use induction on n, since for $n = 1$ the result is trivial. Since X is not linear, there is a point $x \in X$ such that X is not a cone over x. Compare X in $\mathbb{P}^n$ and $p_x(X)$ in $\mathbb{P}^{n-1}$. Then codim $p_x(X) = \text{codim } X - 1$, while $\deg p_x(X) \leq \deg X - 1$ by (5.11). Therefore

$$\deg X \geq \deg p_x(X) + 1$$
$$\geq \text{codim } p_x(X) + 2 \text{ by induction}$$
$$= \text{codim } X + 1. \quad \text{QED}$$

(5.14) **Corollary.** *If $x \in X^r \subset \mathbb{P}^n$ and $y \in \mathbb{P}^n$ satisfies:*
    i) $y \notin X$
    ii) $y \notin E^*_{x,X}$,
    iii) $\overline{yx} \cap X = \{x\}$.

Then let $X' = p_y(X) \subset \mathbb{P}^{n-1}$, $x' = p_y(x) \in X'$. Then

$$\text{mult}_x(X) = \text{mult}_{x'}(X') \cdot \deg(\text{res}_X p_y).$$

Proof. Choose a sufficiently general $M^{n-r-1} \subset \mathbb{P}^{n-1}$ through x', let $L = p_y^{-1}(M) \cup \{y\}$ and compare $X \cap L$ with $X' \cap M$.

(5.15) Corollary. *If $x \in X^r \subset \mathbb{P}^n$, then x is a smooth point of X if and only if $\text{mult}_x(X) = 1$.*

Proof. If x is smooth, then clearly mult is 1. In the example preceding (5.11), we checked the converse for hypersurfaces. The general case can be deduced from the hypersurface case using (5.14). In fact, for all sufficiently general $L^{n-r-1} \subset \mathbb{P}^n$, the following hold:

    i) $L \cap X = \phi$

    ii) $L \cap E^*_{x,X} = \phi$

    iii) $\overline{L,x} \cap X = \{x\}$ (cf. (2.32)).

Then $\text{mult}_x X = 1 \Rightarrow \text{mult}_{p_1(x)}(p_L(X)) = 1 \Rightarrow p_L(x)$ smooth on $p_L(X)$. So if $p_L(X) = V(F_L)$, then the homogeneous equation $F_L \circ p_L$ vanishes on X and has a non-zero differential at x. Varying L, we leave it to the reader to check that we get $n - r$ such functions that vanish on X with independent differentials at x.   QED

Let's look at some examples. Look first at subvarieties $X^r \subset \mathbb{P}^n$ of degree 2. By (5.13), X is a hypersurface in a linear subspace $L^{r+1} \subset \mathbb{P}^n$. So assume $L = \mathbb{P}^n$, $X = V(F)$, F a quadratic form. Now what happens when you project from $x \in \mathbb{P}^n$:

    Case 1. $x \notin X$: then $p_x : X \to \mathbb{P}^{n-1}$ is surjective, of degree 2

    Case 2. $x \in X$, x singular on X. Then $\text{mult}_x X = 2$ and X is a cone over x.

    Case 3. $x \in X$, x smooth on X. Then by (5.11) p_x defines a birational map from X to $\mathbb{P}^{n-1}$. More precisely, we get a diagram:

$$
\begin{array}{ccc}
 & B_x(X) & \\
{\scriptstyle p_1} \swarrow & & \searrow {\scriptstyle p_2} \\
X & & \mathbb{P}^{n-1}
\end{array}
$$

where p_1 and p_2 are surjective and almost everywhere injective. How far is p_2 from being injective?

This is worked out most easily by explicit computation:

Let $(X_i), (Y_i)$ be coordinates in $\mathbb{P}^n$ and $\mathbb{P}^{n-1}$, $X = V(X_0 X_1 + X_2^2 + \ldots + X_n^2)$, $x = (1, 0, \ldots, 0)$, and p_x be the map $Y_i = X_i$, $1 \leq i \leq n$. Then if $y = (y_1, \ldots, y_n) \in \mathbb{P}^{n-1}$, then one finds:

$$
p_2^{-1}(y) = \text{the point}
\begin{cases}
X_0 = -\dfrac{y_2^2 + \ldots + y_n^2}{y_1} \\[2ex]
X_i = y_i, \quad 1 \leq i \leq n
\end{cases}
\quad \text{if } y_1 \neq 0
$$

$$
= \text{the point}
\begin{cases}
X_0 = 1 \\[2ex]
X_i = 0, \quad 1 \leq i \leq n \quad \sum_2^n y_i^2 \neq 0
\end{cases}
\quad \text{if } y_1 = 0
$$

$$\text{any of} = \text{the line} \quad \text{of points} \quad \begin{cases} X_0 = t \\ X_i = sY_i, \quad 1 \le i \le n \end{cases} \quad \text{if } y_1 = \sum_2^n y_i^2 = 0.$$

For instance, we may "draw" the loci of real points in low-dimensional cases like this:

$\boxed{n = 2}$ p_1 and p_2 are bijective.

$\boxed{n = 3}$:

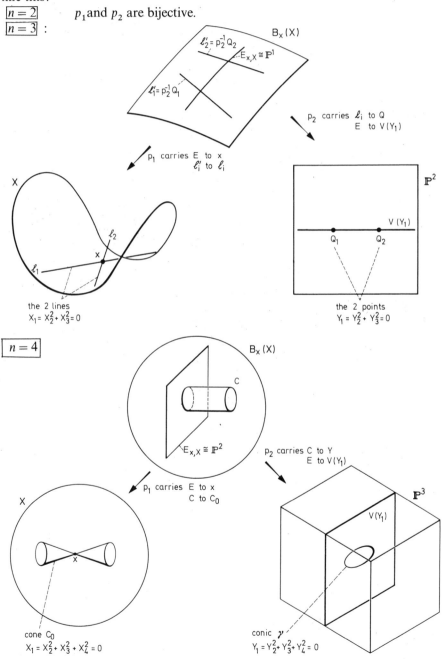

$\boxed{n = 4}$

What subvarieties $X^r \subset \mathbb{P}^n$ are there of degree 3? Assuming that $\mathbb{P}^n$ is the smallest linear space containing X, it follows from (5.13) that either $r = n - 1$, hence $X = V(F)$, F a cubic polynomial, or $r = n - 2$. In the second case, it can be shown that X is defined by the vanishing of the three 2×2-minors of a linear 3×2 matrix:

$$\begin{pmatrix} a_1, a_2, a_3 \\ b_1, b_2, b_3 \end{pmatrix}, \; a_i, b_i \text{ linear forms in } X_0, \ldots, X_n.$$

Assuming this, set $M = V(a_1, \ldots, b_3)$: a linear space of codimension at most 6. Clearly X is a "cone" over M; i.e., if L is a maximal linear space disjoint from M (so dim $L \leq 5$), and $X_0 = X \cap L$, then

$$X = \bigcup_{\substack{x \in M \\ y \in X_0}} \overline{xy}.$$

Now look at X_0 in L. If dim $L \leq 2$, then dim $X_0 \leq 0$ and no such irreducible X_0 exists. There are 3 cases left: dim $L = 3, 4$ or 5. We state without proof what in suitable coordinates X_0 must be:

Case 1. $L \cong \mathbb{P}^3$, $X_0 \cong$ the *twisted cubic curve*

$$= \text{the image of } \mathbb{P}^1 \text{ in } \mathbb{P}^3 \text{ via the map}$$
$$(s,t) \overset{\text{def}}{\longrightarrow} (s^3, s^2t, st^2, t^3)$$
$$= \text{locus where } rk\begin{pmatrix} X_0 & X_1 & X_2 \\ X_1 & X_2 & X_3 \end{pmatrix} = 1.$$

Case 2. $L \cong \mathbb{P}^4$, $X_0 \cong$ the *rational cubic scroll*
$$= \text{image of the embedding of } B_x(\mathbb{P}^2) \text{ in } \mathbb{P}^4, \text{ gotten as the}$$
$$\overset{\text{def}}{\;} \text{closure of the image of}$$
$$\mathbb{P}^2 - (0,0,1) \longrightarrow \mathbb{P}^4$$
$$(x, y, z) \longrightarrow (x^2, xy, y^2, xz, yz)$$
$$= \text{locus where } rk\begin{pmatrix} X_0 & X_1 & X_3 \\ X_1 & X_2 & X_4 \end{pmatrix} = 1.$$

Case 3. $L \cong \mathbb{P}^5$, $X_0 \cong$ image of the Segre embedding
$$\mathbb{P}^1 \times \mathbb{P}^2 \longrightarrow \mathbb{P}^5$$
$$= \text{locus where } rk\begin{pmatrix} X_0 & X_1 & X_2 \\ X_3 & X_4 & X_5 \end{pmatrix} = 1.$$

There is considerable classical literature on lists of subvarieties of low degree, cf. Semple-Roth [1].

§5B. Bezout's Theorem

The most famous theorem of all concerning degree is:

(5.16) **Bezout's Theorem.** *Let X^r, Y^s be 2 subvarieties of $\mathbb{P}^n$. Let $X \cap Y = W_1 \cup \ldots \cup W_k$ be the decomposition of $X \cap Y$ into its irreducible components. Assume:*
   a) dim $W_i = r + s - n$ for all i,

b) *for all i, there is one point $x \in W_i$, (hence in fact for all x in a Zariski-open subset of W_i) such that x is smooth on X and on Y and $T_{x,X}$ and $T_{x,Y}$ intersect transversely in $T_{x,\mathbb{P}^n}$, i.e., $\dim(T_{x,X} \cap T_{x,Y}) = r + s - n$.*
Then

$$\deg X \cdot \deg Y = \sum_{i=1}^{k} \deg W_i.$$

We will summarize properties (a) and (b) (resp. property (a) alone) by saying that X and Y meet transversely (resp. meet properly) Note that if (b) holds at x, then there exist $f_1, \ldots, f_{n-r}$ and $g_1, \ldots, g_{n-s}$ defining X and Y near x with all independent differentials: hence $X \cap Y$ is smooth at x and is defined by the f's and g's together. Thus $x \notin W_j (j \neq i)$, x is smooth on W_i and $T_{x,W_i} = T_{x,X} \cap T_{x,Y}$.

Proof. This is done in 3 steps of which the 1st two are essentially trivial and the 3rd is the key. Some notation: let $W_i' \subset W_i$ denote a Zariski-open set where X and Y intersect as in (b).

Step I. The case $Y^s = L^s$, a linear space. In this case, let $M^{n-r} \subset L^s$ be a sublinear space such that
  a) $M \cap W_i \subset W_i'$
  b) $M \cap W_i$ is a finite set of transversal intersections, i.e., if $x \in M \cap W_i$, then $T_{x,M} \cap T_{x,W_i} = (0)$.
Then we have

$$X \cap M = (X \cap L) \cap M = \bigcup_{i=1}^{k} W_i \cap M.$$

Moreover, for all $x \in W_i \cup M$, by (a) and (b) we have

$$\begin{aligned}
T_{x,X} \cap T_{x,M} &= (T_{x,X} \cap T_{x,Y}) \cap T_{x,M} \\
&= T_{x,W_i} \cap T_{x,M} \\
&= (0)
\end{aligned}$$

hence X and M meet transversely. Therefore

$$\begin{aligned}
\deg X \cdot \deg Y &= \deg X \\
&= \#(X \cap M) \\
&= \sum \#(W_i \cap M) \\
&= \sum \deg W_i.
\end{aligned}$$

Step II. Reduction of the general case to the case $r + s = n$. Start by choosing a linear space L^{2n-r-s} meeting each W_i transversely in points in the open subset W_i'. Then

$$\begin{aligned}
\sum \deg W_i &= \sum \#(L \cap W_i) \\
&= \#(X \cap Y \cap L).
\end{aligned}$$

Now decompose $Y \cap L$ into its components:

$$Y \cap L = Z_1 \cup \ldots \cup Z_l.$$

By (3.28), $\dim Z_i \geq n - r$ for all i. But since $X \cap Z_i$ is a finite non-empty set by

(3.30), dim $Z_i \leq n - r$ for all i. Thus dim $Z_i = n - r$. Now for i, choose a point $x_i \in X \cap Z_i$. Since $x_i \in X \cap Y \cap L$, x_i is in one of the W's: say W_k. Now look at the diagram of tangent spaces:

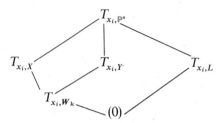

Since X and Y meet transversely at x_i and W_k and L meet transversely at x_i, it follows that $T_{x_i,X} \cap T_{x_i,Y} \cap T_{x_i,L} = (0)$. Therefore $T_{x_i,Y}$ and $T_{x_i,L}$ intersect properly, hence Y and L meet transversely at x_i and $T_{x_i,Y} \cap T_{x_i,L} = T_{x_i,Z_i}$. Moreover $T_{x_i,X} \cap T_{x_i,Z_i} = (0)$ hence X and Z_i meet transversely at x_i. Therefore

$$\begin{aligned}
\#(X \cap Y \cap L) &= \sum \#(X \cap Z_i) \\
&= \sum \deg X \cdot \deg Z_i \quad \text{(by } r + s = n \text{ case of Theorem)} \\
&= \deg X \cdot (\deg Y \cdot \deg L) \quad \text{(by Step I)} \\
&= \deg X \cdot \deg Y.
\end{aligned}$$

Step III. Now assume $r + s = n$. The idea now is to move X continuously in a family $X_t \subset \mathbb{P}^n$, where $X_1 = X$ but in the limit X_0 collapses to a linear space M:

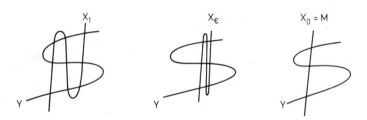

If $d = \deg X$, $e = \deg Y$, then as $X_\varepsilon \to M$, almost all points of M will be the limit of d points of X_ε. But M meets Y in e points. Essentially by the implicit function theorem (in the proof we use (3.26)) we will see that the number of intersections of Y and X_t remains constant as t varies, collapsing in the limit to e points each counted d times.

Now to carry this idea out, first note that projections can be carried out "internally and continuously" in $\mathbb{P}^n$ as well as externally. To be precise, let

$$L^{s-1}, M^r \subset \mathbb{P}^n$$

be any disjoint subspaces and write them

$$L = V(X_0, \ldots, X_r)$$

$$M = V(X_{r+1}, \ldots, X_n)$$

in suitable coordinates. Define

$$\sigma : (\mathbb{P}^n - L) \times \mathbb{C} \longrightarrow \mathbb{P}^n - L$$

by

(*) $$\sigma(x_0, \ldots, x_n, t) = (x_0, \ldots, x_r, tx_{r+1}, \ldots, tx_n)$$

and let $\sigma_t =$ restriction of σ to $(\mathbb{P}^n - L) \times \{t\}$. Then σ_t, for $t \neq 0$, is a projective transformation of $\mathbb{P}^n$ leaving L and M pointwise fixed but gradually pulling all points off L nearer and nearer M as t tends to 0. And σ_0 is the projection $p_L : \mathbb{P}^n - L \to \mathbb{P}^r$ with the image realized as the subspace M of $\mathbb{P}^n$. We will call σ_0 a *linear retraction* of $\mathbb{P}^n - L$ onto M.

Next we choose L and M subject to the following conditions:
a) $L \cap X = \emptyset$
b) M meets Y transversely
c) if $x \in M \cap Y$, then res $\sigma_0 : X \to M$ is smooth over x
[To see that this is possible, proceed as follows: 1st choose M such that (b) holds and that if $M \cap Y = \{y_1, \ldots, y_k\}$, then $y_i \notin X$. 2nd look at the family L_ξ of linear $s - 1$-spaces. Then $L_\xi \cap X = \emptyset$ for ξ in a Zariski open set in the ξ-space. And for each $i, p_{y_i}(L_\xi)$ meets $p_{y_i}(X)$ transversely for ξ in a Zariski-open. If all of these hold, then $\overline{y_i, L_\xi}$ meets X transversely for all i, hence $p_{L_\xi} : X \to M$ is smooth over y_i.]

Now for all $t \in \mathbb{C}$, look at the intersection $\sigma_t(X) \cap Y$. Fit these together into the fibres of a map of algebraic sets as follows: define the algebraic set

$$\mathfrak{X} = \{(x, y, t) \mid \sigma_t x = y\} \subset X \times Y \times \mathbb{C}$$

and consider the restriction of the 3rd projection:

$$\pi : \mathfrak{X} \longrightarrow \mathbb{C}.$$

Note that $\dim(X \times Y \times \mathbb{C}) = n + 1$ and that $\mathfrak{X}$ is defined everywhere locally by n equations:

(**)
$$X_i(\sigma_t x) = X_i(y), \quad 1 \leq i \leq n$$
$$X_1, \ldots, X_n \text{ affine coord. in } \mathbb{P}^n.$$

Therefore, all components of $\mathfrak{X}$ have dimension ≥ 1. Moreover

$$\pi^{-1}(1) \cong X \cap Y, \text{ hence } \# \pi^{-1}(1) = \# (X \cap Y)$$
$$\pi^{-1}(0) \cong \sigma_0^{-1}(Y \cap M), \text{ hence } \# \pi^{-1}(0) = \deg X \cdot \deg Y \text{ by (b) and (c)}.$$

Therefore the Theorem follows from (3.26) provided that π is smooth over 0 and 1.

(A) at 1: Let $x \in X \cap Y$. If you restrict the differentials of the equations (**) defining $\mathfrak{X}$ to $T_{(x,x,1), X \times Y \times (1)} \cong T_{x,X} \oplus T_{x,Y}$, they become the linear maps

$$T_{x,X} \oplus T_{x,Y} \longrightarrow T_{x, \mathbb{P}^n} \longrightarrow \mathbb{C}$$
$$(a, b) \longmapsto a - b$$
$$a \longmapsto dX_i(a)$$

and since $T_{x,X} \cap T_{x,Y} = (0)$, their common kernel is (0). Thus these differentials are all independent and π is smooth at $(x, x, 1)$.

(B) at 0: let $x \in X$ be such that $\sigma_0(x) = y \in M \cap Y$. Choose coordinates $X_1, \ldots, X_n$ at y such that M is given by $X_{r+1} = \ldots = X_n = 0$. Then the equations (**) restrict on $X \times Y \times (0)$ to the equations:

$$X_i \circ \sigma_0(x) = X_i(y), \quad 1 \leq i \leq r$$

$$0 = X_i(y), \quad r+1 \leq i \leq n.$$

Now look at their differentials in $T_{(x,y,0), X \times Y \times (0)} \cong T_{x,x} \oplus T_{y,Y}$. Since M and Y meet transversely at y, the last $n - r = s$ equations have independent differentials on $T_{y,Y}$. And on $T_{x,x} \oplus (0)$, since X and $\sigma_0^{-1}(y)$ meet transversely at x, the first r equations have independent differentials. Therefore their common kernel is (0), so π is smooth at $(x, y, 0)$.   QED

Just as for (5.1), the theorem can be souped up to the case where $\dim W_i = r + s - n$ all i, i.e., X and Y intersect properly, but where X and Y do not intersect transversely. We want to briefly sketch this development. In fact, suppose more generally that X^r, Y^s are subvarieties of any non-singular variety Z^n intersecting properly. Define $\mathscr{Z}_k(Z) =$ free abelian group generated by k-dimensional subvarieties of Z. Then for all components W of $X \cap Y$ one can define a positive integer

$$i(W; X \cap Y)$$

called the *intersection multiplicity* of X and Y along W, and a cycle of dimension $r + s - n$:

$$X \cdot Y = \sum_W i(W; X \cap Y) \cdot W \in \mathscr{Z}_{r+s-n}(Z)$$

called the *intersection product* of X and Y.

Then when $Z = \mathbb{P}^n$ and X and Y intersect properly, Bezout's Theorem states:

$$\deg X \cdot \deg Y = \deg(X \cdot Y)$$

where for any element $\sum n_i Z_i \in \mathscr{Z}_k(\mathbb{P}^n)$, $\deg(\sum n_i Z_i) = \sum n_i \deg Z_i$. In fact, when $Z = \mathbb{P}^n$, one can give a simple *ad hoc* definition of i in terms of mult or its equivalent as follows:

1st case. $r + s = n$. Then define $i(W; X \cap Y) = \# (X \cap \sigma(Y) \cap U)$ where U is a small neighborhood of the point W and $\sigma \in PGL(n+1; \mathbb{C})$ is sufficiently close to the identity.

2nd case. $r + s > n$. Then define $i(W; X \cap Y) = i(W \cap L; X \cap (Y \cap L))$ for a sufficiently general $L^{2n-r-s} \subset \mathbb{P}^n$. Equivalently this is $\# (X \cap \sigma(Y) \cap L \cap U)$.

When i is defined like this, the general case of Bezout's theorem becomes a trivial consequence of the transversal case.

To adapt this definition of i to a general smooth ambient Z, when $r + s = n$ and $W = \{x\}$, one chooses $t_1, \ldots, t_n \in \mathcal{O}_{x,Z}$ local analytic coordinates (i.e., $dt_1, \ldots, dt_n$ independent linear functions on $T_{x,Z}$). Let $U \subset Z$ be a Zariski-open set where the t_i have no poles. Define

$$\psi : (X \cap U) \times (Y \cap U) \longrightarrow \mathbb{C}^n$$

by

$$\psi(x, y) = (t_1(x) - t_1(y), \dots, t_n(x) - t_n(y)),$$

and define

$$i(x; X \cap Y) = \text{mult}_{(x, x)} \psi.$$

When $Z = \mathbb{P}^n$, and the t_i are affine coordinates, the reader can work out what mult ψ means and find that we have our previous definition with σ being a translation in these affine coordinates! In general, it amounts to making a local analytic translation σ in the $\{t_i\}$ coordinate system and counting $\#(X \cap \sigma Y \cap U)$. When $r + s > n$, take a smooth point x on W, assume $t_1, \dots, t_{r+s-n}$ are coordinates on W at x and replace ψ by

$$\psi' : (X \cap U) \times (Y \cap U) \longrightarrow \mathbb{C}^{r+s}$$

$$\psi'(x, y) = (t_1(x), t_1(y), \dots, t_{r+s-n}(x), t_{r+s-n}(y), t_{r+s-n+1}(x) - t_{r+s-n+1}(y), \dots,$$
$$t_n(x) - t_n(y)).$$

The chief obstacle in this approach is to verify that $\text{mult}_{(x,x)} \psi$ is independent of the choice of the $\{t_i\}$. This can be done brutally, but perhaps a better way is by Hilbert-Samuel polynomials (cf. Appendix to Chapter 6). A good reference on the theory of intersection multiplicity is the book of Samuel [1].

§ 5C. Volume of a Projective Variety; Review of Homology, De Rham's Theorem, Varieties as Minimal Submanifolds

Suppose $V^r \subset \mathbb{P}^n$ is a non-singular subvariety. There is a quite natural Riemannian metric on $\mathbb{P}^n$ and a natural question to ask is, in terms of this metric, what is the $2r$-dimensional volume of V^r? The really remarkable fact is that this volume depends only on r and the degree of V. This beautiful theorem is the most elementary and intuitive result in the repertoire of Kähler differential geometry, whose higher developments dominate the whole transcendental theory of varieties, which is why I am including it in this introductory volume.

Let me begin by a few general remarks on Riemannian metrics on complex manifolds. A Riemannian metric on any manifold M is a set of positive definite symmetric real bilinear forms $q_a(x, y)$ on the tangent spaces $T_{a,M}$ to M, varying in a C^∞ way with a. If M is a complex manifold, $T_{a,M}$ is a complex vector space and the most natural real forms q_a to choose from are the real parts of Hermitian forms H. From linear algebra, we have the basic fact:

(5.17) **Lemma.** *Let V be a complex vector space. The following 3 sets are canonically isomorphic:*

(i) *Hermitian forms H on V, i.e., $H : V \times V \to \mathbb{C}$ such that*

$$H(\alpha x_1 + \beta x_2, y) = \alpha H(x_1, y) + \beta H(x_2, y)$$

$$H(y, x) = \overline{H(x, y)}$$

(ii) *quadratic forms q on V invariant under multiplication by i, i.e., $q : V \times V \to \mathbb{R}$ such that*

$$q(\alpha x_1 + \beta x_2, y) = \alpha q(x, y) + \beta q(x_2, y)$$
$$q(y, x) = q(x, y)$$
$$q(ix, iy) = q(x, y)$$

(iii) *exterior 2 forms ω on V invariant under multiplication by i, i.e., $\omega : V \times V \to \mathbb{R}$
such that*

$$\omega(\alpha x_1 + \beta x_2, y) = \alpha \omega(x_1, y) + \beta \omega(x_2, y)$$
$$\omega(y, x) = -\omega(x, y)$$
$$\omega(ix, iy) = \omega(x, y).$$

The isomorphism is set up by:

$$q = \mathrm{Re}\ H$$
$$\omega = \mathrm{Im}\ H$$
$$H(x, y) = q(x, y) - iq(ix, y)$$
$$H(x, y) = \omega(ix, y) + i\omega(x, y).$$

Moreover H is positive definite if and only if q is positive definite.
(Proof left to reader.)

(5.18) **Definition.** *A Hermitian metric on a complex manifold M is a set of positive definite Hermitian forms $H_a(x, y)$ on the tangent spaces $T_{a,M}$ varying in a C^∞ way with a.*

By the lemma, a Hermitian metric $\{H_a\}$ defines a Riemannian metric via $q_a = \mathrm{Re}\ H_a$, and a fundamental 2-form ω by $\omega_a = \mathrm{Im}\ H_a$, both of which are "invariant by i" and in turn characterize H. If $z_1, \ldots, z_n$ are local coordinates in $U \subset M$, then $dz_1, \ldots, dz_n$ are complex-linear functions $T_{a,M} \to \mathbb{C}$ for all $a \in U$ and altogether give a system of coordinates on each vector space $T_{a,M}$. In terms of these, we can expand:

$$H_a(x, y) = \sum_{i,j=1}^{n} g_{ij}(a) dz_i(x) \overline{dz_j(y)}$$

where $g_{ij}(a)$ is a Hermitian matrix, i.e., $\overline{g_{ij}(a)} = g_{ji}(a)$.

(5.19) **Definition.** *A Hermitian metric $\{H_a\}$ is Kähler if the fundamental 2-form ω is closed:*

$$d\omega = 0.$$

To illustrate the above, let's put a metric on $\mathbb{P}^n$. To do this, first embed $\mathbb{P}^n$ differentiably in a vector space as follows:

$$\phi : \mathbb{P}^n \lhook\joinrel\longrightarrow \left\{ \begin{array}{l} \text{real vector space } \mathscr{H} \text{ of} \\ (n+1) \times (n+1) \text{ Hermitian matrices } A \end{array} \right\} \cong \mathbb{R}^{(n+1)^2}$$

$$(X_0, \ldots, X_n) \longmapsto (A_{ij}) = (X_i \bar{X}_j).$$
normalized
     so that
$$\sum |X_i|^2 = 1$$

This makes sense because given 2 sets of homogeneous coordinates $(X_0, \ldots, X_n)$, $(X'_0, \ldots, X'_n)$ such that $X'_i = \lambda X_i$, $\sum |X_i|^2 = 1$ and $\sum |X'_i|^2 = 1$, then $|\lambda| = 1$, hence $X_i \bar{X}_j = X'_i \bar{X}'_j$. Moreover, if $\sigma \in U(n+1, \mathbb{C})$, then ϕ is σ-equivariant in the sense:

$$\phi(\sigma x) = \sigma \cdot \phi(x) \cdot \sigma^{-1}.$$

Now $\mathscr{H}$ has a standard $U(n+1)$-invariant positive definite quadratic form on it:

$$q(A, B) = \text{tr}(A \cdot B).$$

Via the differential of ϕ, we get for all $x \in \mathbb{P}^n$:

$$T_{x, \mathbb{P}^n} \xrightarrow{\quad d\phi \quad} T_{\phi(x), \mathscr{H}} \cong \mathscr{H}$$

hence by pull-back we get a $U(n+1)$-invariant Riemannian metric on $\mathbb{P}^n$. Now $U(n+1)$ acts transitively on $\mathbb{P}^n$, and the matrix

$$\begin{pmatrix} 1 & 0 \\ 0 & iI_n \end{pmatrix}$$

acts as mult. by i in the affine $\mathbb{P}^n - V(X_0)$, hence by mult. by i in $T_{p_0, \mathbb{P}^n}$, $p_0 = (1, 0, \ldots, 0)$. Therefore this Riemannian metric is everywhere invariant under mult. by i, hence is Hermitian. In fact, a rather brutal computation shows that in affine coordinates $x_1 = X_1 / X_0, \ldots, x_n = X_n / X_0$, the metric equals:

$$H = \frac{\sum dx_i \overline{dx}_i}{1 + \sum |x_i|^2} - \frac{(\sum x_i \overline{dx}_i)(\sum \bar{x}_i dx_i)}{(1 + \sum |x_i|^2)^2}.$$

(5.20) **Lemma.** *The above metric H on $\mathbb{P}^n$ is Kähler.*

Proof. Let $\omega = \text{Im } H$ and consider the 3-form $d\omega$. $d\omega$ is also $U(n+1)$-invariant so it suffices to prove that 0 is the only $U(n+1)$-invariant 3-form. Let η be such a 3-form. For all $x \in \mathbb{P}^n$, there is a $\sigma_x \in U(n+1)$ such that $\sigma_x(x) = x$ and $d\sigma_x$ in $T_{x, \mathbb{P}^n}$ is mult. by -1 (if $x = p_0$, let $\sigma_x = \begin{pmatrix} 1 & 0 \\ 0 & -I_n \end{pmatrix}$; in general, take a suitable conjugate of this). Therefore η_x would be a 3-form on $T_{x, \mathbb{P}^n}$ such that

(*) $$\eta_x(-a, -b, -c) = \eta_x(a, b, c).$$

But by linearity,

$$\eta_x(-a, -b, -c) = (-1)^3 \eta_x(a, b, c) = -\eta_x(a, b, c).$$

Therefore $\eta_x = 0$.   QED

Next note that every complex manifold M has a canonical orientation. In fact, if M is a real k-dimensional manifold, by an *orientation* we mean a choice for each $a \in M$ of a positive half-line in the one-dimensional real vector space $\Lambda^k \text{Hom}(T_{a,M}, \mathbb{R})$ of k-forms, varying continuously with a. If M is a complex n-dimensional, hence real $2n$-dimensional, for all $a \in M$, let

$$\omega: \underbrace{T_{a,M} \times \ldots \times T_{a,M}}_{n \times} \longrightarrow \mathbb{C}$$

be a non-zero complex-linear n-form near a. Then it is easy to check that

$$\frac{\omega \wedge \bar{\omega}}{i^n} : \underbrace{T_{a.M} \times \ldots \times T_{a.M}}_{2n \times} \longrightarrow \mathbb{C}$$

in fact takes on real values and is non-zero. If ω' is any other ω, then $\omega = f(a) \cdot \omega'$, with $f(a) \neq 0$, hence

$$\frac{\omega \wedge \bar{\omega}}{i^n} = |f(a)|^2 \cdot \frac{\omega' \wedge \omega'}{i^n}.$$

Thus we have a canonical orientation determined by requiring $\omega \wedge \bar{\omega}/i^n$ to be everywhere positive.

The next step is *Wirtinger's inequality*. This is a lemma out of linear algebra again:

$V = $ complex vector space of complex dimension n
$H = $ pos. def. Herm. form on V; $q = \mathrm{Re}\, H$: $\omega = \mathrm{Im}\, H$
$W \subset V$ a real $2k$-dimensional subspace.

We get on W two things:

a) $\mathrm{res}_W q$, a positive definite quadratic form, hence $\|\mathrm{res}_W q\|$ a volume element on W (i.e., $\in |\Lambda^{2k} \hat{W}|)^*$.

b) $\mathrm{res}_W \omega$, a real 2-form, hence $\mathrm{res}_W \omega^{\wedge k}$, a real $2k$-form, hence $|\mathrm{res}_W \omega^{\wedge k}|$ a 2nd volume element on W.

Then:

(5.21) $\quad \begin{cases} |\mathrm{res}_W \omega^{\wedge k}| \leq \|\mathrm{res}_W q\|, \\ \quad \text{equality if and only if } W \text{ is a complex subspace} \end{cases}$

Proof. Let $\bar{q} = \mathrm{res}_W q$, $\bar{\omega} = \mathrm{res}_W \omega$. First of all, there is a basis $e_1, \ldots, e_k, f_1, \ldots, f_k$ of W such that

a) $q(e_i, e_j) = q(f_i, f_j) = \delta_{ij}, q(e_i, f_j) = 0$

b) $\omega(e_i, e_j) = \omega(f_i, f_j) = 0, \quad \omega(e_i, f_j) = \alpha_i \delta_{ij}.$

This follows for instance from the standard theory of normal forms of skew-symmetric matrices Ω with respect to transformations $\Omega' = A \cdot \Omega \cdot A^{-1}$, A orthogonal. In terms of a set of coordinates $x_1, \ldots, x_k, y_1, \ldots, y_k$ on W dual to the vector e_i, f_i, this says

$$\bar{q} = \sum dx_i^2 + \sum dy_i^2$$
$$\bar{\omega} = \sum \alpha_i dx_i \wedge dy_i.$$

Then

$$\|\bar{q}\| = |dx_1 \wedge \ldots \wedge dy_k|$$
$$|\bar{\omega}^{\wedge k}| = \prod |\alpha_i| \cdot |dx_1 \wedge \ldots \wedge dy_k|.$$

*Strangely enough, after all the recent activity re-doing the foundations of differential geometry and multi-linear algebra by Bourbaki and others, there seems to be no standard notation for the one-dimensional vector space of "volume elements" or Haar measures on a real vector space V. If W is a one-dimensional real vector space, we propose $|W|$ for the associated one-dimensional vector space of elements $\alpha \cdot |w|$, $\alpha \in \mathbb{R}$, $w \in W$, subject to the rule $\alpha |\beta w| = \alpha |\beta| \cdot |w|$. Then the space of volume elements on V is $|\Lambda^n \hat{V}| (\hat{V} = \mathrm{Hom}(V, \mathbb{R}), n = \dim_\mathbb{R} V)$.

But I claim that $|\alpha_i| \le 1$ for all i, and $(|\alpha_i| = 1) \iff (ie_i \in W)$. In fact, define

$$\phi_i : \mathbb{C}^2 \longrightarrow V$$

$$(\alpha, \beta) \longrightarrow \alpha e_i + \beta f_i$$

and consider the Hermitian form $H_i = H \circ (\phi_i \times \phi_i)$ on $\mathbb{C}^2$. Since $H = q + i\omega$, H_i is given by the 2×2 matrix:

$$A_i = \begin{pmatrix} 1 & i\alpha_i \\ -i\alpha_i & 1 \end{pmatrix}.$$

But H_i is always positive semi-definite, hence $\det A_i = 1 - \alpha_i^2 \ge 0$, hence $|\alpha_i| \le 1$. Moreover,

$$|\alpha_i| = 1 \iff H_i \text{ not positive definite}$$
$$\iff \phi_i \text{ not injective}$$
$$\iff ie_i \in \mathbb{R}e_i \oplus \mathbb{R}f_i.$$

But in fact, if ie_i is in W at all, using the fact that $q(ie_i, x) = \mathrm{Re}\, H(ie_i, x) = \mathrm{Im}\, H(e_i, x) = \omega(e_i, x)$, it follows that ie_i is a multiple of f_i. This proves that $|\alpha_i| = 1$ iff $ie_i \in W$.   QED

We now come to the key point:

(5.22) **Theorem.** *Let* $V^r \subset \mathbb{P}^n$ *be a non-singular r-dimensional projective variety. With the Hermitian metric on* $\mathbb{P}^n$ *defined above,*

$$2r - \mathrm{volume}(V) = \deg(V) \cdot 2r - \mathrm{volume}(L)$$

where L is any r-dimensional linear subspace of $\mathbb{P}^n$.

Proof. Let $H = q + i\omega$ denote the Hermitian metric on $\mathbb{P}^n$. Let H_V, q_V, ω_V denote the restrictions to V of these tensors, so that

$$2r - \mathrm{volume}(V) = \int_V \|q_V\|.$$

Since $T_{x,V}$ is a complex subspace of $T_{X,\mathbb{P}^n}$, by Wirtinger's inequality,

$$\int_V \|q_V\| = \int_V |\omega_V^{\wedge r}|$$

$$= \left| \int_V \omega_V^{\wedge R} \right| \quad (\text{since } V \text{ is oriented}).$$

Now use the linear retraction of V onto L as in the proof of Bezout's theorem:

$$\sigma : (\mathbb{P}^n - M) \times \mathbb{C} \longrightarrow (\mathbb{P}^n - M)$$

where $M \cap V = \phi$, $\sigma(x, 1) = x$, $\sigma(x, 0) = p(x)$, p a projection onto L. In particular this gives us by restriction a map

$$\tau : V \times [0,1] \longrightarrow \mathbb{P}^n.$$

Let $\eta = \tau^*(\omega^{\wedge r})$. Then applying Stokes' theorem to the manifold with boundary $V \times [0,1]$,

$$\int_{V \times (1)} \eta - \int_{V \times (0)} \eta = \int_{V \times [0,1]} d\eta = \int_{V \times [0,1]} \tau^*(d\omega^{\wedge r}) = 0$$

since $\mathbb{P}^n$ *is Kähler, hence* $d\omega = 0$! On the other hand, res $\tau : V \times (0) \to \mathbb{P}^n$ is a surjective map from V to L which, outside a lower-dimensional subvariety J of L is a d-to-1 covering ($d = \deg V$). Now the $2r$-measure of J is zero because J is a finite union of analytic manifolds of various dimensions $2k, k < r$. Therefore

$$\int_{V \times (0)} \eta = d \cdot \int_{L} \mathrm{res}_{L}(\omega^{\wedge r}),$$

hence

$$\mathrm{vol}(V) = \left| \int_V \omega_V^{\wedge r} \right| = \left| \int_{V \times (1)} \eta \right| = \left| \int_{V \times (0)} \eta \right| = d \cdot \left| \int_L \omega_L^{\wedge r} \right| = d \cdot \mathrm{vol}(L). \qquad \text{QED}$$

The idea of the above proof ties in very closely with the DeRham theorem and if we use this theorem we can strengthen Theorem (5.22) to characterize algebraic varieties as *minimal submanifolds* of $\mathbb{P}^n$. To explain this clearly, we would like next to give an outline of the DeRham theory and show precisely what it means for $\mathbb{P}^n$. Good references for what we need are Warner [1] and Singer-Thorpe [1]. Start with an arbitrary real C^∞-manifold M. First some definitions:

(5.23) The singular homology groups. $H_k(M, \mathbb{Z})$. Define these as follows:

$$\Delta^k = \begin{cases} \text{the convex hull in } \mathbb{R}^{k+1} \text{ of the unit vectors} \\ (0, \ldots, 1, \ldots, 0), \end{cases}$$

$$\varepsilon_i : \Delta^{k-1} \longrightarrow \Delta^k = \begin{cases} \text{the linear map taking the vertices} \\ v_0, \ldots, v_{k-1} \text{ of } \Delta^{k-1} \text{ to the vertices} \\ v_0, \ldots, \hat{v}_i, \ldots, v_k \text{ of } \Delta^k, \end{cases}$$

$$C_k(M) = \begin{cases} \text{group of linear combinations } \sum n_i f_i, n_i \in \mathbb{Z}, \\ f_i \text{ continuous maps from } \Delta^k \text{ to } M, \end{cases}$$

$$\partial : C_k(M) \longrightarrow C_{k-1}(M) = \begin{cases} \text{the linear map defined on the generators } f \text{ of } C_k(M) \\ \text{by } \partial f = \sum_{i=0}^{k} (-1)^i f \circ \varepsilon_i, \end{cases}$$

$$H_k(M, \mathbb{Z}) = \ker[\partial : C_k(M) \longrightarrow C_{k-1}(M)] / \mathrm{Im}[\partial : C_{k+1}(M) \longrightarrow C_k(M)].$$

(5.24) Differentiable chains. Since Δ^k is a manifold with boundary and M is a manifold, we can talk of the subgroup:

$$C_k(M_{\mathrm{diff}}) \subset C_k(M)$$

generated by the C^∞ maps $f : \Delta^k \to M$. As before, we can form ker $\partial / \mathrm{Im}\ \partial$ on these smaller chain groups and we get a group and a map

$$H_k(M_{\text{diff}}, \mathbb{Z}) \longrightarrow H_k(M, \mathbb{Z}).$$

(5.25) **Triangulations.** Every C^∞-manifold is triangulable—i.e., there exists for all $l, 0 \le l \le \dim M$, a countable set K_l of C^∞ embeddings $f: \Delta^l \to M$ such that:

a) $M =$ disjoint union of the images $f(\text{Int } \Delta^l), f \in K_l$

b) $f \in K_l \Rightarrow f \circ \varepsilon_i \in K_{l-1}$ all i

c) $\forall\ x \in M$, only a finite number of simplices $f(\Delta^l)$ contain x. For a proof, cf. Munkres [1]. Given a triangulation K, there is the even smaller chain group:

$$C_k(K) \subset C_k(M_{\text{diff}})$$

generated by the maps $f \in K_k$. As before we can form ker $\partial/\text{Im } \partial$ on these smaller groups and we get a group and a map

$$H_k(K, \mathbb{Z}) \longrightarrow H_k(M_{\text{diff}}, \mathbb{Z}).$$

(5.26) **Theorem.** $H_k(K, \mathbb{Z}) \xrightarrow{\ \approx\ } H_k(M_{\text{diff}}, \mathbb{Z}) \xrightarrow{\ \approx\ } H_k(M, \mathbb{Z}).$

We assume this basic fact. The isomorphism of the 1st and 3rd groups is proven in every book on algebraic topology. The isomorphism of the 2nd group with the others is handled by the usual method of acyclic carriers*, (cf. Warner [1]).

(5.27) **De Rham cohomology groups.** let

$\Omega^k(M) =$ real vector space of C^∞ k-forms on M,

$d: \Omega^k(M) \to \Omega^{k+1}(M) =$ usual exterior differentiation,

$H^k_{DR}(M) = \text{Ker}[d: \Omega^k(M) \to \Omega^{k+1}(M)]/\text{Im}[d: \Omega^{k-1}(M) \to \Omega^k(M)]$

$\qquad = $ closed k-forms/exact k-forms.

Now if we use the construction (II.) of H_k by *differentiable* chains, we get a pairing:

a) $\forall\ \omega \in \Omega^k(M), \forall\ \sigma = \sum n_i f_i \in C_k(M_{\text{diff}}),$

let
$$\langle \omega, \sigma \rangle = \sum n_i \int_{\Delta^k} f_i^* \omega$$

b) then $\langle d\omega, \sigma \rangle = \langle \omega, \partial \sigma \rangle$ (this is Stokes' theorem!)

c) hence if $d\omega = 0, \partial\sigma = 0, \langle \omega, \sigma \rangle$ depends only on ω mod exact forms and on σ mod boundaries. Thus we get a *period pairing*:

$$H^k_{DR}(M) \times H_k(M_{\text{diff}}, \mathbb{Z}) \longrightarrow \mathbb{R}.$$

(5.28) **Theorem of De Rham.** *Via this pairing, $H^k_{DR}(M)$ is isomorphic to* $\text{Hom}(H_k(M, \mathbb{Z}), \mathbb{R})$. *Hence, going backwards, if $H_k(M, \mathbb{R}) \underset{\text{def}}{=} H_k(M, \mathbb{Z}) \otimes \mathbb{R}$ is finite-dimensional, it is the dual of $H^k_{DR}(M)$.*

We assume this too. For a proof cf. Singer-Thorpe [1] and Warner [1]. There

---

*i.e., 1) introduce a covering $\{U_\alpha\}$ such that all intersections $U_S = U_{\alpha_1} \cap \ldots \cap U_{\alpha_k}$ are diffeomorphic to convex subsets of $\mathbb{R}^n$; 2) let $C^*(M_{\text{diff}}), C^*(M)$ be the subcomplexes generated by simplices σ such that Im $\sigma \subset U_\alpha$, some α; 3) show that $H_k(C^*(M_{\text{diff}})) \simeq H_k(C(M_{\text{diff}}))$ and $H_k(C^*(M)) \simeq H_k(C(M))$; show that for all $(\alpha_1, \ldots, \alpha_k), H_k(C(U_{S,\text{diff}})) \simeq H_k(C(U_S)) \simeq (0)$ or $\mathbb{Z}$(if $k > 0$ or $k = 0$); 5) show that the complex $C^*(M)/C^*(M_{\text{diff}})$ with carriers $C^*(U_S)/C^*(U_{S,\text{diff}})$ is acyclic by the lemma of acyclic carriers (e.g., Greenberg [1], p. 148).

is still another very important definition to make before we can use the De Rham theory effectively:

(5.29) **Fundamental classes.** Let $N \subset M$ be a compact oriented connected k-dimensional submanifold of M. In standard advanced calculus books, one defines $\int_N \omega (\omega \in \Omega^k(M))$ by coordinate charts and a partition of unity on N (cf. for instance, Lang [2], Chs. 17–18; or Spivak [1]). We would like to know that this is a special case of our period pairing, i.e., that there is a differentiable chain $\sum n_i \sigma_i$ on M such that

$$\sum n_i \int_{\Delta^k} \sigma_i^* \omega = \int_N \omega, \qquad \text{all } \omega \in \Omega^k(M).$$

In fact, let $\{K_l\}_{0 \leq l \leq k}$ be a triangulation of N. Then I claim:

(5.30) **Lemma.** i) *The group of k-cycles $v = \sum_{\sigma \in K_k} n(\sigma) \cdot \sigma$ is infinite cyclic, generated by a "fundamental cycle" v_1, for which $n(\sigma) = \pm 1$, all $\sigma \in K_k$.*
   ii) *There is a sign $\varepsilon = \pm 1$ such that*

$$\int_{v_1} \omega = \varepsilon \int_N \omega, \qquad \text{all } \omega \in \Omega^k(N).$$

Proof. We shall show first that if $\partial(\sum n(\sigma)\sigma) = 0$ and for one simplex $\sigma_0 \in K_k$, $n(\sigma_0) = 0$, then $n(\sigma) = 0$ for all σ. In fact, since N is a manifold every $\tau \in K_{k-1}$ is a face of exactly 2 k-simplices $\sigma', \sigma'' \in K_k$. Call σ', σ'' *adjacent*. The fact that the coefficient of τ in $\partial(\sum n(\sigma)\sigma)$ is 0 means precisely that $n(\sigma') = \pm n(\sigma'')$, the sign depending on which of the faces of σ' and σ'' τ is. Since N is connected, any two k-simplices of K_k are connected through a finite chain of adjacent pairs. Therefore in fact $|n(\sigma)|$ is independent of σ. Thus $n(\sigma_0) = 0 \Rightarrow n(\sigma) = 0$ all σ. This proves that *either* $H_k(K, \mathbb{Z}) = (0)$ *or* $H_k(K, \mathbb{Z}) \cong \mathbb{Z}$ and its 2 generators are cycles of type $\sum n(\sigma) \cdot \sigma, |n(\sigma)| = 1$ all σ. Next, if $S = \bigcup_{\substack{\sigma \in K_l \\ l < k}} \sigma(\text{Int } \Delta^l)$, then S has measure 0 on N, hence for all $\omega \in \Omega^k(N)$,

$$\int_N \omega = \int_{N-S} \omega$$

$$= \sum_{\sigma \in K_k} \int_{\sigma(\text{Int } \Delta^k)} \omega$$

$$= \sum_{\sigma \in K_k} n(\sigma) \cdot \int_{\Delta^k} \sigma^* \omega$$

$$= \left\langle \sum_{\sigma \in K_k} n(\sigma) \cdot \sigma, \omega \right\rangle$$

where $n(\sigma) = \pm 1$, the sign depending on whether the map σ is orientation preserv-
ing or not. In particular, if $\omega = d\eta$, then since N is a compact manifold with no
boundary $\int_N d\eta = 0$ by Stokes' theorem, hence

$$0 = \left\langle \sum_{\sigma \in K_k} n(\sigma) \cdot \sigma, d\eta \right\rangle$$

$$= \left\langle \partial \left(\sum_{\sigma \in K_k} n(\sigma)\sigma \right), \eta \right\rangle$$

hence $\sum n(\sigma)\sigma$ is a cycle v and $\int_N \omega = \int_v \omega$ for all ω.   QED

(5.31) **Corollary.** $H_k(N, \mathbb{Z}) \cong \mathbb{Z}$ and the 2 generators correspond to the 2 possible
orientations on N.

(5.32) **Definition.** If N is given an orientation, let $[N]$ be the image in $H_k(M, \mathbb{Z})$
via the natural homomorphism $H_k(N, \mathbb{Z}) \to H_k(M, \mathbb{Z})$ of the positive generator of
$H_k(N, \mathbb{Z})$. $[N]$ is called the fundamental class of the oriented N.

We can now return to complex projective space $\mathbb{P}^n$ and compute everything
explicitly in this case:

(5.33) **Theorem.** (i) If $0 \le k \le n$, then $H_{2k}(\mathbb{P}^n, \mathbb{Z}) \cong \mathbb{Z}$, the fundamental class $[L^k]$
of a k-dimensional linear space being a generator,
    (ii) $H_k(\mathbb{P}^n, \mathbb{Z}) = (0)$ if k is odd or $k > 2n$,
    (iii) the 2k-form $\omega^{\wedge k}$ is a basis of the space of closed 2k-forms on $\mathbb{P}^n$ mod
exact 2k-forms,
    (iv) if k is odd, all closed k-forms are exact.

Proof. Computing $H_k(\mathbb{P}^n, \mathbb{Z})$ by a triangulation, we see that $H_k = (0)$ ff $k > 2n$.
Moreover by (5.31), $H_{2n}(\mathbb{P}^n, \mathbb{Z}) \cong \mathbb{Z}$ and is generated by $[\mathbb{P}^n]$. To compute $H_k(\mathbb{P}^n, \mathbb{Z})$
for $k < 2n$, we use induction on n and a linear retraction. Let $x \in \mathbb{P}^n$ and let

$$\tau : \mathbb{P}^n - \{x\} \times [0, 1] \longrightarrow \mathbb{P}^n - \{x\}.$$

$$\tau(y, 1) = y$$

$$\tau(y, 0) = p(y) \in L^{n-1}$$

be the retraction of $\mathbb{P}^n - (x)$ onto L. We make use of the basic:

(5.34) **Lemma.** If X and Y are manifolds and $F : X \times [0, 1] \to Y$ is a continuous
map, let $F_t(x) = F(x, t)$. Then for every singular cycle σ on X, $F_0 \circ \sigma - F_1 \circ \sigma$ is a
boundary on Y.

(Cf. Greenberg [1], Ch. 11). Now if σ_1 is any k-cycle on $\mathbb{P}^n$, $k < 2n$, first triangulate $\mathbb{P}^n$ so that x is in the interior of one of the 2n-cells. Then σ_1 is homologous
to a cycle σ_2 consisting of k-cells in this triangulation, hence in $\mathbb{P}^n - \{x\}$. Applying
the lemma to τ, σ_2 is homologous to $\sigma_3 = p \circ \sigma_2$, which is a cycle on L. This means
that the natural map:

$$H_k(L,\mathbb{Z}) \longrightarrow H_k(\mathbb{P}^n,\mathbb{Z})$$

is surjective. Since $L \cong \mathbb{P}^{n-1}$, this proves by induction that $H_k(\mathbb{P}^n,\mathbb{Z}) = (0)$ if k is odd and that $H_{2k}(\mathbb{P}^n,\mathbb{Z})$ is generated by $[L^k]$-although $[L^k]$ might still be zero or of finite order in $H_{2k}(\mathbb{P}^n,\mathbb{Z})$. But $\omega^{\wedge k}$ is a closed $2k$-form and

$$\int_{L^k} \omega^{\wedge k} = \text{volume}\,(L) > 0,$$

hence $[L^k]$ is of infinite order. This proves (i) and (ii) and it also shows that $\omega^{\wedge k}$ is not exact, hence (iii) and (iv) follow from De Rham's theorem.   QED

It is now possible to define the degree of any compact oriented $2k$-dimensional submanifold N in $\mathbb{P}^n$. In fact the fundamental class $[N]$ will equal $d \cdot [L^k]$ for some $d \in \mathbb{Z}$ and if N is an algebraic variety, then this d is just the degree of N by (5.22) and De Rham's Theorem. In general, we call this d the *degree* of N. We can now prove the following beautiful complement to (5.22):

(5.35) **Theorem.** *Let $N \subset \mathbb{P}^n$ be a compact oriented $2k$-dimensional submanifold of degree d. Then*

$$2k\text{-volume}\,(N) \geq |\deg N| \cdot 2k\text{-volume}\,(L^k),$$

with equality if and only if N is an algebraic subvariety.

Proof. We mimic the proof of (5.22). In the notation of that proof,

$$2k\text{-vol}(N) = \int_N \|q_N\|$$

$$\geq \int_N |\omega_N^{\wedge k}| \qquad \text{by Wirtinger's inequality}$$

$$\geq \left| \int_N \omega_N^{\wedge k} \right|$$

$$= |\langle [N], \omega^{\wedge k} \rangle|$$

$$= |\deg N| \cdot \langle [L^k], \omega^{\wedge k} \rangle$$

$$= |\deg N| \cdot 2k\text{-vol}(L^k).$$

Say equality holds. Then in fact, for every $x \in N$, the 2 volume elements $\|q_N\|$ and $|\omega^{\wedge k}|$ must be equal on $T_{x,N}$. By Wirtinger's inequality again this means that $T_{x,N}$ must be a complex subspace of $T_{x,\mathbb{P}^n}$. Let $X_1,\dots X_n$ be affine coordinates near x. Then this means first of all that for some set of k of these coordinates $X_{i_1},\dots,X_{i_k}$, projection onto $\mathbb{R}^{2k}$ by the real and imaginary parts of these coordinates has a non-zero jacobian. Therefore N can be described locally near x as the graph of a set of C^∞ complex-valued functions:

$$X_l = f_l(\text{Re}\,X_{i_1}, \text{Im}\,X_{i_1}, \dots, \text{Re}\,X_{i_k}, \text{Im}\,X_{i_k}), l \notin \{i_1,\dots,i_k\}.$$

But now it is easy to see that the fact that $T_{x,N}$ is a *complex* subspace of $T_{x,\mathbb{P}^n}$ means precisely that each f_l equals a complex-linear function of $X_{i_1},\ldots,X_{i_k}$ plus a function vanishing to 2nd order at x; which is the same as saying that each f_l satisfies the Cauchy-Riemann equations (4.10) at x! Therefore the f_l's are all analytic. This proves that N is a complex submanifold of $\mathbb{P}^n$, hence by Chow's theorem N is an algebraic subvariety too.   QED

The theory of volumes of analytic sets can be pushed further in several ways: first one can show that an analytic set has locally finite volume even near singular points, and then extend (5.22) to singular subvarieties; second one can look at analytic sets X in $\mathbb{C}^n$ and study the growth of the volume of $S \cap$ (ball of radius r) with r, obtaining a characterization of affine varieties S and a new proof of Chow's theorem. A good reference for these questions is Stolzenberg [1]. Even more important is the further investigation of the implications of the Kähler condition, $d\omega = 0$, for the complex analytic refinements of the De Rham theory (e.g., the Hodge decomposition of $H^n_{DR}(X)$ into $\sum_{p+q=n} H^{p,q}(X)$) and for integral geometry: cf. Weil [1] and Cornalba-Griffiths article in A.M.S. [1]. Concerning further results on the topology of $\mathbb{P}^n$ and its subvarieties, we would like to make reference in passing to Thom's extremely elegant application of Morse theory to prove the following fundamental result of Lefschetz —

a) for $\mathbb{P}^n$ itself, we have the simple fact that $\mathbb{P}^n$ is constructed by attaching* a single $2n$-cell to $\mathbb{P}^{n-1}$,

b) for any non-singular r-dimensional subvariety $X \subset \mathbb{P}^n$, if $H \subset \mathbb{P}^n$ is a hyperplane meeting X transversely, then X is homotopic to the space obtained by attaching a sequence of k-cells *with $r \leq k \leq 2r$*, to $X \cap H$.

This shows for instance that up to dimension $r - 2$, X and $X \cap H$ have the same homology and homology groups. These can both be proven easily by considering the C^∞ real-valued function

$$f(x_0,\ldots,x_n) = \frac{|\sum a_i x_i|^2}{\sum |x_i|^2}$$

where H has equation $\sum a_i x_i = 0$

on $\mathbb{P}^n$ and applying the (easy) Morse theory to this function. For $\mathbb{P}^n$ itself, f takes its minimum 0 on H, has a non-degenerate maximum at $(a_0, a_1, \ldots, a_n)$ and no other critical points. Thus (a) follows from this plus Morse theory. Restricting f to X, it can be shown that f takes its minimum 0 on $X \cap H$ and has otherwise non-degenerate critical points *of index between r and $2r$*. Thus (b) follows. For details on this, the reader is referred to Milnor [1], §3, and Bott [1], §2.

---

*By definition, attaching a k-cell to X means forming the union $\Delta^k \cup X$ modulo identification of points of $\partial \Delta^k$ with points of X via an "attaching map" $f: \partial \Delta^k \to X$.

Chapter 6. Linear Systems

§ 6A. The Correspondence between Linear Systems and Rational Maps, Examples; Complete Linear Systems are Finite-Dimensional

One of the main activities of classical geometry was to find non-trivial birational correspondences between seemingly unrelated projective varieties. For instance, we saw in Chapter 5 that, by projecting from a smooth point, we get a birational correspondence from any irreducible quadric to projective space itself. To find these correspondences, the principle method was by using linear systems. In this section we shall explain this method. For simplicity, we stick to a *smooth r-dimensional variety* X. As in §1C and 2A, we define the group of divisors:

$$\text{Div}(X) = \text{free abelian group on subvarieties } Z \subset X \text{ of codimension 1}$$

and for all $f \in \mathbb{C}(X), f \neq 0$, we can define the divisor $(f) \in \text{Div}(X)$ of zeroes and poles of f. Recall that $(f) = 0$ if and only if f is a constant, i.e., we have an exact sequence:

$$1 \longrightarrow \mathbb{C}^* \longrightarrow \mathbb{C}(X)^* \longrightarrow \text{Div}(X)$$
$$f \longmapsto (f).$$

(6.1) **Definition.** *2 divisors D_1, D_2 are linearly equivalent, or $D_1 \equiv D_2$ if they differ by a principal divisor: $D_1 - D_2 = (f)$ for some $f \in \mathbb{C}(X)^*$.*

(6.2) **Definition.** *If $D = \sum\limits_{i=1}^{t} n_i Z_i \in \text{Div}(X)$, then*

$$\mathscr{L}(D) = \left\{ f \in \mathbb{C}(X) \middle| \begin{array}{l} f = 0 \text{ or } f \neq 0 \text{ and } (f) + D \geq 0 \\ \text{i.e., all coeff. of } (f) + D \text{ are non-negative} \end{array} \right\}$$

$$= \left\{ f \in \mathbb{C}(X) \middle| \begin{array}{l} \text{ord}_{Z_i}(f) \geq -n_i, 1 \leq i \leq t, \\ \text{and ord}_Z(f) \geq 0, \text{ all other } Z \end{array} \right\}$$

Note that this is a vector space: it is the set of rational functions on X with no poles except at the Z_i and at each Z_i either with poles of at most a certain bounded order, or with zeroes of at least a certain order. We may say that the *fundamental problem of the additive theory of functions on X is to compute* dim $\mathscr{L}(D)$ *for all D.* We will see below that at least dim $\mathscr{L}(D) < +\infty$.

(6.3) **Definition.** *If $D \in \text{Div}(X)$ and $\mathscr{L}(D) \neq (0)$, then we say that $|D|$ exists and let*

$$|D| = \text{the set of non-negative divisors } (f) + D.$$

Thus $|D|$ is the intersection of the coset $D + (\text{princ. divisors})$ with the semigroup of non-negative divisors. Note that

$$(f_1) + D = (f_2) + D$$

if and only if $f_1 = \alpha f_2$, some $\alpha \in \mathbb{C}^*$, hence $|D|$ is isomorphic to the projective space of 1-dimensional subspaces of $\mathscr{L}(D)$. Let dim $|D| = $ (dim of this proj. space) $=$ dim $\mathscr{L}(D) - 1$.

(6.4) **Definition.** *A linear system is a subset L of some $|D|$ such that*

$$V = \{f \in \mathbb{C}(X) | f = 0 \quad or \quad f \neq 0 \text{ and } D + (f) \in L\}$$

is a vector space over $\mathbb{C}$. Equivalently, L is a linear subspace of $|D|$ in its structure of projective space. The linear systems $|D|$ themselves are called complete linear systems. The base points of a linear system L are defined by:

$$(\text{Base pts. of } L) = \bigcap_{D \in L} (\text{Support of } D).$$

If $D_1, \ldots, D_k$ are linearly equivalent effective divisors so that $|D_1| = |D_2| = \ldots = |D_k|$, the linear system L spanned by them is the smallest linear subspace $L \subset |D_1|$ containing all the D_i.

Besides their significance for the additive function theory of X, linear systems derive their great importance from the fact that they give geometric meaning to the homogeneous coordinate ring as follows:

(6.5) **Definition.** *For all $f \in \mathbb{C}[X_0, \ldots, X_n]/I(X), f \neq 0$, and f homogeneous of degree d, define $(f) \in \text{Div}(X)$:*
  let $U_i = $ Zariski open set $X - X \cap (\text{hyperplane } X_i = 0)$
  look at the divisor (f/X_i^d) in U_i
  In $U_i \cap U_j$, f/X_i^d and f/X_j^d differ by a nowhere 0,
  nowhere ∞ function $(X_i/X_j)^d$, so $(f/X_i^d) = (f/X_j^d)$
  let $(f) = $ (the divisor on X equal to (f/X_i^d) in U_i).
Alternately, consider the intersection of X and the hypersurface $H = V(f)$ in $\mathbb{P}^n$:
  Let $X \cap H = Z_1 \cup \ldots \cup Z_t$: then codim $Z_i = 1$
  Assign each Z_i a multiplicity n_i by:
  Choosing $k(i)$ such that $X_{k(i)} \not\equiv 0$ on Z_i,
  Setting $n_i = \text{ord}_{Z_i}(f/X_{k(i)}^d)$
  Then $\sum n_i Z_i$ is denoted by $X \cdot H$ or by (f), and is called the hypersurface section of X by H.

(6.6) **Definition.** *For all $d \geq 1$, let $L_X(d)$ equal the set of divisors (f), where $f \in [\mathbb{C}[X_0, \ldots, X_n]/I(X)]_d$. Since for all $f, g \in [\mathbb{C}[X_1, \ldots, X_n]/I(X)]_d$ $(f) = (g) + (f/g)$, it follows that $L_X(d)$ is a linear system and is called the "linear system cut out on X by the hypersurfaces of degree d."*

But we can go much further with this construction—say $Z : X \to \mathbb{P}^m$ is *any* rational map and assume that $Z[X] \not\subset$ any hyperplane. Then I claim that the inverse images of the hyperplanes, $Z^{-1}[H]$, form a linear system L on X if they are assigned suitable multiplicities.

Let $F = \{x \in X | \dim Z[x] \geq 1\}$: then codim $F \geq 2$ and $Z : (X - F) \to \mathbb{P}^m$ is regular.

Let $Y_0, \ldots, Y_m$ be coordinates in $\mathbb{P}^m$ and let H be the hyperplane $l = \sum \alpha_i Y_i = 0$.

Everywhere on $\mathbb{P}^m$, H is defined locally as the zeroes of $l/Y_i = 0$, some i.

So everywhere on $X - F, Z^{-1}(H)$ is defined locally as the zeroes of the regular function $(l/Y_i) \circ Z$ (for some i).

Define $Z^*(H)$ to be the divisor on X which on $X - F$ is defined as $(l/Y_i \circ Z)$ locally.

Note that if $H_1 - H_2 = (f)$ on $\mathbb{P}^m$, then $Z^*(H_1) - Z^*(H_2) = (f \circ Z)$ where $\circ Z$ is the induced map $\mathbb{C}(\mathbb{P}^m) \to \mathbb{C}(X)$. From this it follows immediately that the set of divisors $Z^*(H)$ is a linear system L_Z on X. Now the set of hyperplanes $H \subset \mathbb{P}^m$ forms a projective space since it equals the set of linear forms $\sum \alpha_i Y_i$ modulo scalars: this is called the dual projective space $\check{\mathbb{P}}^m$. And the set-theoretic isomorphism $H \mapsto Z^*(H)$ of $\check{\mathbb{P}}^m$ with L_Z is an isomorphism of projective spaces. The Italians used the phrase: "the map Z refers the linear system L_Z to the system of hyperplanes" to describe this situation. Moreover, I claim that

(6.7) **Lemma.** (a) $F = (set\ of\ base\ points\ of\ L_Z)$
      (b) Supp $Z^*(H) = Z^{-1}[H]$.

Proof. In fact, by definition Supp $Z^*(H)$ and $Z^{-1}[H]$ agree outside F. And if $x \notin F$, then $Z[x] =$ one point; so $\exists H$ such that $H \cap Z[x] = \phi$, or in other words $x \notin Z^{-1}[H]$. Therefore the linear system L_Z has no base points outside F. While if $x \in F$, then $Z[x]$ contains a curve; so $\forall H, H \cap Z[x] \neq \phi$, or $x \in Z^{-1}[H]$. Therefore $F \subset Z^{-1}[H]$ all H. This proves (a). But if $H = V(l)$ is any hyperplane, then

$$(Y_i/l \circ Z) = Z^*(H_i) - Z^*(H) \qquad (\text{where } H_i = V(Y_i))$$

hence on the open set $X - \text{Supp } Z^*(H)$, the functions $f_i = Y_i/l \circ Z$ have no poles. Therefore, over $X - \text{Supp } Z^*(H)$, the subset Z of $X \times \mathbb{P}^m$ is contained in the locus of zeroes of $f_i Y_j - f_j Y_i (0 \leq i,j \leq m)$. Now $\sum a_i f_i = 1$, so for all $x \in X - \text{Supp } Z^*(H), f_i(x) \neq 0$ for some i. Therefore the equations $f_i(x) \cdot Y_j - f_j(x) \cdot Y_i = 0$ determine only one point of $\mathbb{P}^m$, namely $(f_0(x), \ldots, f_m(x))$. Thus Z is single-valued on $X - \text{Supp } Z^*(H)$, so we must have $F \subset \text{Supp } Z^*(H)$. QED

The fundamental result is that conversely, L_Z determines Z up to composition with a projective transformation. More precisely:

(6.8) **Theorem.** *There is a one-one correspondence between*
      a) *linear systems L of dimension m whose base points have codimension ≥ 2, plus projective isomorphisms ϕ of L with $\check{\mathbb{P}}^m$,*
      b) *rational maps $Z: X \to \mathbb{P}^m$ such that $Z[X] \not\subset$ any hyperplane.*
      If $\beta: \mathbb{P}^m \xrightarrow{\approx} \mathbb{P}^m$ is any projective transformation, then in this correspondence, keeping the same L but replacing ϕ by $\beta^{t-1} \circ \phi$ means changing Z to $\beta \circ Z$.

Proof. In one direction, Z determines L_Z and $Z^*: \mathbb{P}^m \xrightarrow{\approx} L_Z$. Let's define this correspondence in the other direction. Starting with L and ϕ, let $D_0 \in L$ be the divisor such that $\phi(D_0)$ has homogeneous coordinates $(1,0,\ldots,0)$, i.e., corresponds to the hyperplane $X_0 = 0$. Then L is the set of 1-dimensional subspaces of a vector space V of functions such that $(f) + D_0$ is effective; and $\check{\mathbb{P}}^m$ is the set of 1-dimensional subspaces of the vector space of linear forms $\sum \alpha_i X_i$. So ϕ is defined by an isomorphism Φ of V with this space of linear forms, and in this, 1 goes over to X_0. Let $f_i = \Phi^{-1}(X_i)$. $1 \leq i \leq m$. Define a regular map

$$Z_0 : X - \operatorname{Supp} D_0 \longrightarrow \mathbb{P}^m$$

by

$$Z_0(x) = (1, f_1(x), \ldots, f_m(x)).$$

Closing up the graph of Z_0 in $X \times \mathbb{P}^m$, we get a rational map Z. We leave it to the reader to check that these constructions are inverse to each other and behave as asserted when ϕ is changed. QED

(6.9) **Definition.** *The L defined by Z in (6.8) will be denoted L_Z; any of the Z's defined by L in (6.8) will be denoted Z_L.*

Let's see how we can relate these new concepts to ideas already encountered and compute Z_L in some cases. We will omit details, as these are all more or less formal manipulations:

A) Assume $X \subset \mathbb{P}^n, X \not\subset$ hyperplane. Then the linear system L_1 of hyperplane sections of X defines the given embedding of X in $\mathbb{P}^n$. The rational map defined by a *sub*-linear system $L'_1 \subset L_1$ satisfying the base point condition is the projection

$$p_M : X \longrightarrow \mathbb{P}^m$$

where the center M of the projection is the intersection of the hyperplanes defining divisors in L'_1. For instance, if L'_1 has codimension one in L_1, then we get a projection from a point:

$$p_x : X \longrightarrow \mathbb{P}^{n-1}.$$

(If, however, X is contained in various hyperplanes, let $\mathbb{P}^m \subset \mathbb{P}^n$ be the smallest linear space containing X. Then L_1 defines the given embedding of X in $\mathbb{P}^m$.)

B) More generally, if $L_1 \subset L_2$ are 2 linear systems on X of dimensions n_1, n_2, then it is easy to check that Z_{L_1}, Z_{L_2} are related by the diagram:

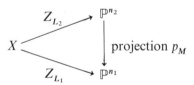

where $M \subset \mathbb{P}^{n_2}$ is the linear space which is the intersection of the hyperplanes corresponding to divisors in L_1. More precisely, if $F = $ set of base points of L_1, then $Z_{L_2}[X - F] \subset \mathbb{P}^{n_2} - M$ and the diagram of true maps

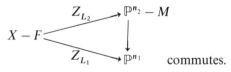

commutes.

C) If $X = \mathbb{P}^n$, all the Z_L's are very easy to write down. In fact, every non-negative divisor D on $\mathbb{P}^n$ equals (f) for some homogeneous polynomial f; hence every linear system L is a subsystem of the linear system L_d of all hypersurfaces of degree d, and the L_d are the only complete linear systems. In general:

$$L = \{(f)\,|\,f \in V, f \neq 0\} \qquad \text{where}$$

V = a vector space of homogeneous f's of degree d.

$$Z_L : \mathbb{P}^n \longrightarrow \mathbb{P}^m$$

is just the map defined by:

$$(x_0,\ldots,x_n) \longrightarrow (f_0(x),\ldots,f_m(x))$$

outside $F = \{x\,|\,f_i(x) = 0, \text{ all } i\}$ of codim ≥ 2. Here are some examples:

C_1) If on $\mathbb{P}^1$, we look at the linear system of all divisors of degree 2 (resp. 3), i.e., the set of all (f), $f(X_0,X_1)$ homogeneous of degree 2 (resp. 3), we obtain a regular correspondence

$$Z : \mathbb{P}^1 \longrightarrow \mathbb{P}^2 \quad (\text{resp. } \mathbb{P}^1 \longrightarrow \mathbb{P}^3),$$

whose image is a non-singular conic, (resp. the twisted cubic of (§5A)).

C_2) The Cremona transformation $T : (x_0,x_1,x_2) \longrightarrow \left(\dfrac{1}{x_0},\dfrac{1}{x_1},\dfrac{1}{x_2}\right)$ discussed in (2.20) is also the map:

$$(x_0,x_1,x_2) \longrightarrow (x_1 x_2, x_0 x_2, x_0 x_2, x_0 x_1)$$

and the vector space of forms $\alpha x_1 x_2 + \beta x_0 x_2 + \gamma x_0 x_1$ is the vector space of all quadratic forms $f(x_0,x_1,x_2)$ such that $f(0,0,1) = f(0,1,0) = f(1,0,0) = 0$. Therefore T is defined by the linear system of all conics with the 3 base points $(0,0,1),(0,1,0)$, and $(1,0,0)$.

C_3) On the other hand, the 5-dimensional linear system of *all* conics in $\mathbb{P}^2$ defines a regular correspondence

$$Z : \mathbb{P}^2 \longrightarrow \mathbb{P}^5$$

whose image F has not been discussed yet. Taking a basis of monomials for the set of quadratic f's, we find that F is the locus of points:

$$(x^2, xy, xz, y^2, yz, z^2), \quad (x,y,z) \in \mathbb{P}^2.$$

It is called the *Veronese surface* and turns out to be smooth of degree 4. It has the remarkable property is that all its secants $\overline{xy}, x, y \in F$, lie on a hypersurface of $\mathbb{P}^5$, which, if you think about it, is one dimension lower than you'd expect.

D) If $X \subset \mathbb{P}^n$, then the rational map $Z_d : X \to \mathbb{P}^{N(d)}$ defined by the linear system L_d of hypersurface sections is just the following map —

    a) embed $\mathbb{P}^n$ itself in $\mathbb{P}^{N'(d)}$ by the complete linear system of hypersurfaces of degree d — call this i,

    b) let $\mathbb{P}^{N(d)}$ be the smallest linear subspace of $\mathbb{P}^{N'(d)}$ containing $i(X)$,

    c) then Z_d is equal to

$$\text{res } i : X \longrightarrow \mathbb{P}^{N\,(d)}.$$

One can easily check that via i, $\mathbb{P}^n$ is in *biregular* correspondence with its image $i(X)$, hence X is also in biregular correspondence with $Z_d(X)$. $Z_d(X)$ is called the *d-tuple embedding of X*.

More generally, if $X \subset Y$ are smooth varieties and L is a linear system on Y whose base points B_Y are of codimension ≥ 2 and don't contain X, we can form a rational map by closing up the graph of the regular map

$$(X - X \cap B_Y) \subset (Y - B_Y) \xrightarrow{\;Z_L\;} \mathbb{P}^m.$$

What linear system defines this? This is readily seen to be the following "reduced trace" of L:

a) first form $\mathrm{tr}_X L$ by taking all divisors $D \in L$ such that $X \nsubseteq \mathrm{Supp}(D)$, assigning multiplicities to the components of $X \cap D$ as in (6.5) above, thus giving a divisor $X \cdot D$ on X, and finally

$$\mathrm{tr}_X L = \{X \cdot D \mid D \in L, X \nsubseteq \mathrm{Supp}\, D\}$$

b) take the greatest effective divisor D_0 which is contained in all divisors of $\mathrm{tr}_X L$ and setting

$$\mathrm{Red}\ \mathrm{Tr}_X L = \{D - D_0 \mid D \in \mathrm{Tr}_X L\}.$$

E) The Segre embedding $\mathbb{P}^n \times \mathbb{P}^m \to \mathbb{P}^{nm+n+m}$ is the map given by the linear system on $\mathbb{P}^n \times \mathbb{P}^m$ of divisors of "bidegree $(1,1)$", i.e., defined by bi-homogeneous equations of degree 1 in each set of variables:

$$V\left(\sum_{i=0}^{n} \sum_{j=0}^{n} a_{ij} X_i Y_j \right).$$

Alternately, we may describe this linear system as the span of the divisors

$$H_1 \times \mathbb{P}^m + \mathbb{P}^n \times H_2$$
$$H_1 \subset \mathbb{P}^n, H_2 \subset \mathbb{P}^m \text{ hyperplanes.}$$

F) Putting (D) and (E) together, we get a new definition of $B_x(X)$. Recall that $B_x(X)$ is defined to be the closure in $\mathbb{P}^n \times \mathbb{P}^{n-1}$ of the points $(y, p_x(y))$, $y \in X - \{x\}$. Via the Segre embedding, we can embed $\mathbb{P}^n \times \mathbb{P}^{n-1}$ in $\mathbb{P}^{n^2+n-1}$: call this s. Let $\mathbb{P}^m$ be the smallest linear subspace containing $s(B_x(X))$. Thus we have a projective embedding $B_x(X) \longrightarrow \mathbb{P}^m$ and a rational map

$$Z : X \longrightarrow s(B_x(X)) \subset \mathbb{P}^m.$$

Now s is defined by the linear system which is the span of the divisors $(H_1 \times \mathbb{P}^{n-1} + \mathbb{P}^n \times H_2)$: and the pull-backs of $H_1 \times \mathbb{P}^{n-1}$, resp. $\mathbb{P}^n \times H_2$, by

$$(X - \{x\}) \longrightarrow B_x(X) \subset \mathbb{P}^n \times \mathbb{P}^{n-1}$$

are just the hyperplane sections of $X - \{x\}$, resp. its hyperplane sections by hyperplanes through x. The span of the pairwise sum of these is just the linear system of all quadratic hypersurface sections $X \cdot V(F)$ given by quadratic forms F with $F(x) = 0$. Thus we conclude that Z is defined by the codimension 1 subsystem L'_2 of L_2 of quadric sections through x, and $B_x(X)$ may be defined as the image of this rational map.

So far we have not used the concept of *complete* linear systems. We now prove:

(6.10) **Theorem.** *Let $X^r \subset \mathbb{P}^n$ be a smooth projective variety. Then the linear system $L_X(d)$ of hypersurface sections is complete if d is sufficiently large.*

Proof. Replacing $\mathbb{P}^n$ by a smaller linear space, we can assume that $X \not\subset$ any hyperplane. Let $H_i = (X_i) \in L_X(1), 0 \le i \le n$. Suppose $\overline{L_X(d)}$ is the complete linear system containing $L_X(d)$ and that $D \in \overline{L_X(d)}$. Then $D \equiv dH_i$, all i, so we can find $f_i \in \mathbb{C}(X)$ such that

$$(f_i) = D - dH_i.$$

Therefore
$$\begin{aligned}(f_i/f_j) &= (D - dH_i) - (D - dH_j) \\ &= d(H_j - H_i) \\ &= (X_j^d/X_i^d).\end{aligned}$$

Changing the f_i's by suitable constants, we can assume that

$$f_i/f_j = X_j^d/X_i^d \qquad \text{for all } i,j,$$

i.e., $\quad F = X_i^d f_i \in (\text{field of fractions of } \mathbb{C}[X_0,\dots,X_n]/I(X))$

is independent of i. Now note that f_i has no poles in the affine piece $X - X \cap H_i$ of X. Therefore the function f_i is in the affine coordinate ring of $X - X \cap H_i$ by (1.32). Therefore

$$f_i = F_i/X_i^N, \text{ some } F_i \in \mathbb{C}[X_0,\dots,X_n]/I(X) \text{ of degree } N$$

or:

$$X_i^{N-d} \cdot F = X_i^N \cdot f_i = F_i \in \mathbb{C}[X_0,\dots,X_n]/I(X).$$

We now apply to F the second part of the following Proposition, to deduce that $D = (F) \in L_d$ as required.

(6.11) **Proposition.** 1) *Let M be a finitely generated torsion-free graded $\mathbb{C}[Y_0,\dots,Y_r]$-module: i.e., $M = \bigoplus\limits_{k=k_0}^{\infty} M_k$ and for all f homogeneous of degree d, $f \cdot M_k \subset M_{k+d}$. Let $K = \mathbb{C}(Y_0,\dots,Y_r), M_K =$ the localization* of M with respect to K. Then there is a d_0 such that if $d \ge d_0$*

$$\left.\begin{array}{c} m \in M_K \\ Y_0^N \cdot m,\dots, Y_r^N \cdot m \in M_{N+d} \end{array}\right\} \Longrightarrow m \in M_d.$$

2) *Let $R = \mathbb{C}[X_0,\dots,X_n]/\mathfrak{P}$, $\mathfrak{P}$ a homogeneous prime ideal. Then there is a d_0 such that if $d \ge d_0$*

$$\left.\begin{array}{c} f \in \text{fraction field of } R \\ X_0^N f,\dots, X_n^N f \in R_{N+d} \end{array}\right\} \Longrightarrow f \in R_d.$$

Proof. Note that (1) $\Rightarrow$ (2) immediately by applying Noether Normalization (2.29). In fact, by (2.29), we can choose

---

*Recall that, by definition, M_K is the K-vector space of elements $\frac{m}{a}, m \in M, a \in \mathbb{C}[Y_0,\dots,Y_r], a \ne 0$; where $m/a = m'/a'$ if and only if $a'm = am'$.

$$\mathbb{C}[Y_0,\ldots,Y_r]\subset R, \qquad Y_i = \text{linear comb. of } X_0,\ldots,X_n$$

so that R is a finitely generated torsion-free graded $\mathbb{C}[Y_0,\ldots,Y_r]$-module. Then (1) with $M = R$ implies (2). As for (1), one first checks that any such M is a graded submodule of a free module M' on generators $m'_1,\ldots,m'_l$ of suitable degrees:

$$M \subset M' = \bigoplus_{i=1}^{l} \mathbb{C}[Y_0,\ldots,Y_r]\cdot m'_i.$$

(Let $m_1,\ldots,m_k$ be homogeneous generators of M. Then they span the K-vector space M_K, so we may assume $m_1,\ldots,m_l$ are a K-basis of M_K. Then for all $i > l$

$$f_i m_i = \sum_{j=1}^{l} a_{ij} m_j ; f_i, a_{ij}\in\mathbb{C}[Y_0,\ldots,Y_r], \text{ homogeneous.}$$

Let $f = \prod_{i=l+1}^{k} f_i$ and let $m'_i = m_i/f$.) Then note that if $m\in M_K$ and $Y_i^N\cdot m\in M$, at least it follows that $m\in M'$. In fact, expand

$$m = \sum_{i=1}^{l} a_i m'_i.$$

Then $a_i\in K$ and $b_{ij} = Y_j^N a_i\in\mathbb{C}[Y_0,\ldots,Y_r]$. Then for all* $k\neq j$, $Y_k^N b_{ij} = Y_j^N b_{ik}$ so by the UFD property Y_j^N divides b_{ij} in $\mathbb{C}[Y_0,\ldots,Y_r]$, i.e., $a_i\in\mathbb{C}[Y_0,\ldots,Y_r]$. Thus $m\in M'$. Let

$$\tilde{M} = \{m\in M' \mid \exists\, N \text{ such that } Y_i^N m\in M, \text{ all } i\}.$$

Then $\tilde{M}$, being a submodule of M', is finitely generated. If $\tilde{m}_i$ are generators of degree d_i, and $Y_i^N \tilde{m}_j\in M$, all i,j, then

$$(Y_0,\ldots,Y_r)^{N(r+1)}\cdot\tilde{m}_i \subset M$$

(since any monomial in the Y's of degree $N(r+1)$ contains one of the Y_i to at least the Nth power), hence

$$(Y_0,\ldots,Y_r)^{N(r+1)}\tilde{M}\subset M,$$

hence $\tilde{M}_d = M_d$ if $d\geq(\max d_i) + N(r+1)$.   QED

(6.12) **Corollary.** *Every complete linear system is finite-dimensional.*

Proof. Let $D = \sum n_i Z_i$ be any positive divisor on X. Let F_i be a homogeneous polynomial that vanishes on Z_i but not everywhere on X. Let $d_i = \deg F_i$. Then we get

$$G = \prod F_i^{n_i}, \text{ of degree } e = \sum n_i d_i,$$

such that $(G) = D + D_1, D_1$ effective.

Raising G to a power if necessary we can assume that e is large enough so that $L_X(e)$ is complete. Then we find

---

*This argument assumes $r\geq 1$. If $r = 0$, $\mathbb{C}[Y_0]$ is a P.I.D., and we leave it to the reader to find a simple direct argument.

$$\mathscr{L}(D) \subseteq \mathscr{L}((G))$$

hence
$$|D| + D_1 \subseteq |(G)| = L_X(e).$$

It follows that
$$\dim |D| \leq \dim L_X(e) < +\infty. \quad \text{QED}$$

§6B. Differential Forms, Canonical Divisors and Branch Loci

Among all linear systems on a variety X, a key role is played by the so-called *canonical* and *pluricanonical* linear systems. The point of this section is merely to define these, together with the necessary background on differential forms. We begin with a general discussion of differential forms. To put the situation in some perspective, let's contrast 3 cases—

1) if X is a C^∞ differential manifold, $x \in X$, then a real C^∞ *l*-form ω near x is a family of real-linear forms $\omega(x')$:

$$\omega(x') : T_{x',X} \longrightarrow \mathbb{R}, \quad x' \text{ near } x,$$

varying in a C^∞ way with x'. Thus if $\dim X = n$ and $t_1, \ldots, t_n$ are local coordinates near x, we can expand

$$\omega(x') = \sum_{k=1}^n a_k(x') dt_k$$

a_k C^∞ real-valued functions on X

and ω may be thought of naively as a "tensor", an n-tuple $\{a_k\}$ of functions depending on the choice of coordinates and transforming by multiplication by the Jacobian matrix $\{\partial t_k/\partial s_l\}$ when you change from coordinates $\{t_1, \ldots, t_n\}$ to $\{s_1, \ldots, s_n\}$. We may also look at complex C^∞ 1-forms ω. Here

$$\omega(x') : T_{x',X} \longrightarrow \mathbb{C}$$

is real-linear, and

$$\omega(x') = \sum a_k(x') dt_k$$

a_i C^∞ complex-valued functions on X.

2) if X is a complex analytic manifold, $x \in X$, then among all the C^∞ complex-valued forms, we have the so-called (1,0)-forms. Here, since $T_{x',X}$ has a complex structure we require that

$$\omega(x') : T_{x',X} \longrightarrow \mathbb{C}$$

is complex-linear. If $n = $ complex dimension of X, $z_1, \ldots, z_n$ are local analytic coordinates, hence $u_k = \text{Re } z_k, v_k = \text{Im } z_k$ are real coordinates, then a general 1-form

$$\omega(x') = \sum_{k=1}^n a_k(x') du_k + b_k(x') dv_k$$

is a (1,0)-form if $b_k(x') = i \cdot a_k(x')$, i.e.,

$$\omega(x') = \sum_{k=1}^{n} a_k(x')(du_k + idv_k) = \sum_{k=1}^{n} a_k(x')dz_k.$$

Furthermore, among all these (1,0)-forms, one has the analytic 1-forms ω: this means that the coefficients $a_k(x')$ are required to be analytic functions of x'. This requirement is independent of the choice of coordinates $\{z_k\}$, because if $\{w_k\}$ is a 2nd choice, then

$$\omega = \sum_k a_k dz_k = \sum_k a_k \sum_l \frac{\partial z_k}{\partial w_l} \cdot dw_l = \sum_l \left(\sum_k a_k \frac{\partial z_k}{\partial w_l} \right) dw_l$$

and

$$a_k \text{ analytic} \Longrightarrow \sum_k a_k \frac{\partial z_k}{\partial w_l} \text{ analytic}$$

because the partial derivatives $\partial z_k/\partial w_l$ are analytic.

3) finally suppose $X \subset \mathbb{P}^N$ is a smooth algebraic variety. Then among all sets of local coordinates at x, there are those which are just given by an n-tuple of linear functions on an affine neighborhood of x. Using such coordinates, suppose we require that the coefficients $a_k(x')$ be *regular algebraic functions near x*, i.e., $a_k \in \mathcal{O}_{x,X}$. The ω's with this property are called the *regular algebraic 1-forms*. This condition is independent of the choice of local coordinates by virtue of:

(6.13) **Lemma.** *Let $z_1, \ldots, z_n$ be any local algebraic coordinates, i.e., elements of $\mathcal{O}_{x,X}$ such that, equivalently, $dz_1, \ldots, dz_n$ are independent linear forms on $T_{x,X}$ or that $z_1, \ldots, z_n$ are local analytic coordinates near x. Then for all $f \in \mathcal{O}_{x,X}$, it follows that*

$$\partial f/\partial z_i \in \mathcal{O}_{x,X}, \quad 1 \leq i \leq n.$$

Proof. Let $z_1, \ldots, z_n$ define a regular map $F: U \to \mathbb{C}^n$ on some Zariski-neighborhood of x. Since the z_i are local coordinates at x, $F(U) \supset$ (neigh. of 0), hence $F(U)$ is Zariski-dense. Therefore F extends to a surjective rational map from X to $\mathbb{P}^n$, and F^* gives an injection

$$\mathbb{C}(z_1, \ldots, z_n) \overset{F^*}{\longleftrightarrow} \mathbb{C}(X).$$

Since $\text{tr.d.}_{\mathbb{C}}\mathbb{C}(X) = n$, $\mathbb{C}(X)$ is a finite algebraic extension of $\mathbb{C}(z_1, \ldots, z_n)$. Thus any $f \in \mathcal{O}_{x,X}$ satisfies an equation:

$$a_0(z_1, \ldots, z_n)f^d + a_1(z_1, \ldots, z_n)f^{d-1} + \ldots + a_d(z_1, \ldots, z_n) = 0$$

$a_i \in \mathbb{C}[z_1, \ldots, z_n]$. Applying $\partial/\partial z_i$, it follows that:

$$(da_0 f^{d-1} + (d-1)a_1 f^{d-2} + \ldots + a_{d-1}) \frac{\partial f}{\partial z_i} + \left(\frac{\partial a_0}{\partial z_i} f^d + \frac{\partial a_1}{\partial z_i} f^{d-1} + \ldots + \frac{\partial a_d}{\partial z_i} \right) = 0$$

hence

$$\frac{\partial f}{\partial z_i} = - \frac{\dfrac{\partial a_0}{\partial z_i} f^d + \dfrac{\partial a_1}{\partial z_i} f^{d-1} + \ldots + \dfrac{\partial a_d}{\partial z_i}}{da_0 f^{d-1} + (d-1)a_1 f^{d-2} + \ldots + a_{d-1}} \in \mathbb{C}(X).$$

Therefore $\dfrac{\partial f}{\partial z_i}$ is both a rational function on X and an analytic function with no poles near x. Therefore, by (1.32E), $\dfrac{\partial f}{\partial z_i} \in \mathcal{O}_{x,X}$.   QED

One should contrast this lemma with fact, from the general theory of fields, that since $\text{tr.d.}_{\mathbb{C}}\mathbb{C}(X) = n$, then

$$n = \left\{ \begin{array}{l} \text{dim. of vector space of derivations} \\ D : \mathbb{C}(X) \to \mathbb{C}(X), \text{ with } D \equiv 0 \text{ on } \mathbb{C} \\ \text{—call this } \text{Der}_{\mathbb{C}}\mathbb{C}(X) \end{array} \right\}$$

and if $z_1, \ldots, z_n \in \mathbb{C}(X)$ is any transcendence basis, then the derivations $\partial/\partial z_i$ of $\mathbb{C}(z_1, \ldots, z_n)$ extend uniquely to derivations still written $\partial/\partial z_i$ of $\mathbb{C}(X)$ which form a basis of the above vector space $\text{Der}_{\mathbb{C}}\mathbb{C}(X)$. The above lemma says that if, in addition, $z_1, \ldots, z_n$ are local algebraic coordinates, then each derivation $\partial/\partial z_i$ maps $\mathcal{O}_{x,X}$ into itself. This can also be proven by purely algebraic methods.

Note that an algebraic 1-form ω, regular at x, is automatically regular on a whole Zariski neighborhood of x: in fact, if $\omega = \sum a_i \cdot dz_i$, then $\{z_1, \ldots, z_n\}$ will be algebraic coordinates and $\{a_1, \ldots, a_n\}$ will have no poles in a Zariski neighborhood of U. Define a *rational 1-form* ω to be an algebraic 1-form ω regular on some non-empty Zariski-open $U \subset X$. If $z_1, \ldots, z_n$ is any transcendence base of $\mathbb{C}(X)$ over $\mathbb{C}$, then ω can be expanded

$$\omega = \sum a_i \cdot dz, \qquad a_i \in \mathbb{C}(X).$$

Naively ω may be thought of as a tensor, i.e. an n-tuple $\{a_i\}$ of rational functions depending on the choice of transcendence base and transforming by multiplication by the Jacobian matrix $\{\partial z_i / \partial w_j\}$ when you change from transcendence base $\{z_i\}$ to $\{w_i\}$.

All these constructions are functorial in straightforward ways: i.e.,

1) $\forall\, f : X \to Y$ C^∞ map of C^∞-manifolds, and real C^∞ 1-forms ω on Y, we get a real C^∞ 1-form $f^*\omega$ on X.

2) $\forall\, f : X \to Y$ analytic map of complex manifolds and analytic 1-forms ω on Y, we get an analytic 1-form $f^*\omega$ on X, and

3) $\forall\, f : X \to Y$ rational map of smooth varieties, regular at x and rational 1-form ω on Y regular at $f(x)$, we get a rational 1-form $f^*\omega$ on X regular at x.

Moreover, all this works equally well in the differentiable, analytic and algebraic cases for higher tensors too, by simply replacing *linear* maps on $T_{x,X}$ by multi-linear ones with various kinds of symmetry. In particular, alternating forms on k variables in $T_{x,X}$ are called "k-forms" and we have in the C^∞-category the usual exterior derivative $\omega \to d\omega$ associating a $(k+1)$-form to a k-form. The key point is that if ω is analytic, then so is $d\omega$ and if ω is algebraic, then so is $d\omega$. This follows from lemma (6.13) again. Note that if X is a complex n-dimensional manifold, then n-forms are objects:

$$\omega = a \cdot dz_1 \wedge \ldots \wedge dz_n$$

and there are no k-forms for $k > n$. Thus all analytic n-forms are closed under d. This generalizes the well-known fact from one-complex variable theory that an analytic 1-form

$$a(z) \cdot dz$$

is closed, which, by applying Stoke's theorem, leads us to Cauchy's theorem. We will discuss this in §7C in connection with residues. In particular, if $n = \dim X$, X a variety, we are especially interested in rational n-forms:

$$\omega(x') : \Lambda^n(T_{x',X}) \longrightarrow \mathbb{C}, \qquad x' \in U, \quad U \text{ Zariski-open in } X$$

given by:

$$\omega = a \cdot dz_1 \wedge \ldots \wedge dz_n, \qquad a \in \mathbb{C}(X), z_1, \ldots, z_n \text{ a transcendence basis of } \mathbb{C}(X).$$

Changing the transcendence base $\{z_i\}$ then means multiplying the coefficient a by the Jacobian determinant $\det(\partial z_i / \partial w_j)$. The set of these forms is a 1-dimensional vector space over $\mathbb{C}(X)$.

We may also look at k-fold n-forms on X:

$$\omega(x') : \Lambda^n(T_{x',X})^{\otimes k} \longrightarrow \mathbb{C}$$

$$\omega = a(dz_1 \wedge \ldots \wedge dz_n)^k$$

which naively are simply elements $a \in \mathbb{C}(X)$ transforming by the kth power of the jacobian determinant. Note that a k_1-fold n-form ω_1 and a k_2-fold n-form ω_2 can be multiplied to form a $(k_1 + k_2)$-fold n-form $\omega_1 \otimes \omega_2$. Thinking of elements of $\mathbb{C}(X)$ as o-fold k-forms, this generalizes scalar multiplication of functions and forms. These forms have a nicely behaved *divisor of zero and poles* $(\omega) \in \mathrm{Div}(X)$. We define this as follows:

for all $W \subset X$ of codimension 1,
choose $x \in W$
choose local coordinates $z_1, \ldots, z_n$ at x
expand $\omega = a \cdot (dz_1 \wedge \ldots \wedge dz_n)^k$
let $\mathrm{ord}_W \omega = \mathrm{ord}_W a$
let $\quad (\omega) = \sum_W \mathrm{ord}_W \omega \cdot W.$

Then for instance we get
i) $(f\omega) = (f) + (\omega), \quad$ all $\ f \in \mathbb{C}(X),$
and more generally, $(\omega_1 \otimes \omega_2) = (\omega_1) + (\omega_2)$ if ω_i is a k_i-fold n-form.
ii) ω regular at x iff $\mathrm{ord}_W \omega \geq 0$, all W through x; ω everywhere regular iff $(\omega) \geq 0.$

(6.14) **Definition.** *The linear equivalence class of divisors (ω) where ω is a non-zero rational n-form on X is called the canonical divisor class. Any member of this divisor class will be denoted by the letter K_X.*

Note that if ω is a k-fold n-form, $(\omega) \in kK_X$.

(6.15) **Definition.** *For all divisors D on X:*

$$\Omega_{k,X}(D) = \left\{ \begin{array}{l} \text{vector space of rational } k\text{-fold } n\text{-forms } \omega \text{ on } X \\ \text{with } (\omega) + D \geq 0 \end{array} \right\}$$

$$\Omega_{k,X} = \ \Omega_{k,X}(0) = \{\text{vector space of } \omega\text{'s with no poles}\}.$$

Moreover, $P_k(X) = \dim \Omega_{k,X}$ are called the kth plurigenera of X, $p_g(X) = P_1(X)$ is called the geometric genus of X and $\bigoplus\limits_{k=0}^{\infty} \Omega_{k,X}$ is called the canonical ring of X. Also $\Omega_X(D)$ (without any k) will denote $\Omega_{1,X}(D)$.

Note that in terms of K_X, we may write:

$$\Omega_{k,X}(D) \cong \mathscr{L}(kK_X + D)$$

$$\Omega_{k,X} \cong \mathscr{L}(kK_X).$$

This is somewhat sloppy but useful. It means: fix any non-zero rational n-form ω. Write K_X for (ω). Then we get an isomorphism

$$\mathscr{L}(kK_X + D) \xrightarrow{\;\approx\;} \Omega_{k,X}(D)$$

$$f \longrightarrow f \cdot \omega^{\otimes k}$$

because $(f \cdot \omega^{\otimes k}) = kK_X + (f)$, hence $(f \cdot \omega^{\otimes k}) + D \geq 0$ if and only if $(f) + (kK_X + D) \geq 0$. The great importance of the plurigenera and the canonical ring is that if X_1, X_2 are smooth birationally equivalent varieties, their plurigenera are equal and their canonical rings is isomorphic. This follows from:

(6.16) Proposition. *Let X_1, X_2 be smooth projective varieties and $Z \subset X_1 \times X_2$ a birational correspondence between them. Then Z induces an isomorphism*

$$Z^* : \Omega_{k,X_1} \xrightarrow{\;\approx\;} \Omega_{k,X_2}.$$

Proof. We have seen in Chapter 3 that there are fundamental sets $F_i \subset X_i$ such that

    a) $Z : X_1 - F_1 \longrightarrow X_2$
    b) $Z^{-1} : X_2 - F_2 \longrightarrow X_1$

are regular and codimension$_{X_i}(F_i) \geq 2$. Given $\omega \in \Omega_{k,X_2}$, by the regular map in (a), we get a k-fold n-form $Z^*(\omega)$ on $X_1 - F_1$. But then $Z^*(\omega)$ is a rational form on X_1 with poles only in F_1 which has codimension at least 2. Thus $Z^*(\omega)$ is everywhere regular, i.e., $Z^*(\omega) \in \Omega_{k,X_1}$. Applying the same argument with the regular map in (b), we find $(Z^{-1})^*$ maps Ω_{k,X_1} to Ω_{k,X_2}. These maps are clearly inverses. QED

Let's calculate $K_{\mathbb{P}^n}$. Recall from §2A that $\mathrm{Pic}(\mathbb{P}^n) \cong \mathbb{Z}$ via degree. It is usual to let H stand for the divisor class of hyperplanes on $\mathbb{P}^n$: then any divisor D on $\mathbb{P}^n$ is linearly equivalent to kH, some $k \in \mathbb{Z}$. In fact

(6.17) Lemma. $K_{\mathbb{P}^n} \equiv -(n+1)H$.

Proof. Let $x_1 = X_1/X_0, \ldots, x_n = X_n/X_0$ be affine coordinates in $\mathbb{P}^n - V(X_0)$. Let $\omega = dx_1 \wedge \ldots \wedge dx_n$. Since $\{x_i\}$ are analytic coordinates in $\mathbb{P}^n - V(X_0)$, the coefficient a of ω is just 1 for these coordinates, so ω has no zeroes or poles in $\mathbb{P}^n - V(X_0)$. Thus $(\omega) = kV(X_0)$ for some $k \in \mathbb{Z}$. Let $y_0 = X_0/X_n$, $y_1 = X_1/X_n, \ldots$, $y_{n-1} = X_{n-1}/X_n$ be affine coordinates in $\mathbb{P}^n - V(X_n)$. Then

$$y_0 = 1/x_n,\, y_1 = x_1/x_n, \ldots, y_{n-1} = x_{n-1}/x_n,$$

and one checks that

$$\omega = dx_1 \wedge \dots \wedge dx_n = \frac{(-1)^n}{y_0^{n+1}} dy_0 \wedge dy_1 \wedge \dots \wedge dy_{n-1}.$$

Therefore $(\omega) = -(n+1)V(X_0)$.   QED

Since $|kH| = \phi$ whenever $k < 0$, it follows that $|kK_{\mathbb{P}^n}| = \phi$ and $\Omega_{k,\mathbb{P}^n} = (0)$ whenever $k \geq 1$. Thus the canonical ring is not very interesting on $\mathbb{P}^n$! This is closely related to the fact that from differential-geometry $\mathbb{P}^n$ stands at one end of the curvature spectrum, having the most positive curvature of any algebraic variety. And the canonical class K_X turns out to represent minus the cohomology class of the Ricci curvature of X (cf. Cornalba-Griffiths [A.M.S. 1]). But it is still hoped that in some sense in the "majority" of cases, the canonical ring carries significant information about X. In dimensions 1 and 2, this turns out to be true (cf. Mumford [2] and Bombieri-Husemoller [A.M.S. 1]).

There is another much more elementary sense in which all projective varieties have more positive canonical class than $\mathbb{P}^n$ itself, and this is the basis of the classical, purely geometric definition of the canonical class K_X. To explain this, we first define the branch locus:

(6.18) **Definition.** *Given a regular dominating map $f : X^r \to Y^r$ of smooth r-dimensional varieties, define an effective divisor B on X^r, called the branch locus, as follows:*
for all $x \in X^r$, let $y_1, \dots, y_r$ be local algebraic coordinates on Y near $f(x)$.
Near x, define B to be $(f^*(dy_1 \wedge \dots \wedge dy_r))$.

Note that this makes sense because if $y_1', \dots, y_r'$ are any other local algebraic coordinates on Y near $f(x)$, then

$$dy_1' \wedge \dots \wedge dy_r' = \det(\partial y_i'/\partial y_j) \cdot dy_1 \wedge \dots \wedge dy_r$$

and the jacobian determinant is a unit in $\mathcal{O}_{y,Y}$. Thus

$$f^*(dy_1' \wedge \dots \wedge dy_r') = u \cdot f^*(dy_1 \wedge \dots \wedge dy_r), \ u \text{ unit in } \mathcal{O}_{x,X},$$

hence in a neighborhood of x,

$$(f^*(dy_1' \wedge \dots \wedge dy_r')) = (f^*(dy_1 \wedge \dots \wedge dy_r)).$$

Explicitly, if $x_1, \dots, x_r$ are local algebraic coordinates at x, then near x:

$$B = \text{divisor of the jacobian determinant } \det(\partial(f^*y_i)/\partial x_j).$$

Clearly:

$$\text{Supp}(B) = \{x \in X \,|\, f \text{ not smooth at } x\},$$

i.e., B is just the set of non-smooth points of f with suitable multiplicities assigned.

(6.19) **Proposition.** *Given a smooth variety $X^r \subset \mathbb{P}^n$, let M^{n-r-1} be a linear space disjoint from X^r and let $p_M : X^r \to \mathbb{P}^r$ be the projection. Then if $B_M = $ branch locus of p_M, and H is a hyperplane section of X, we have*

$$K_X \equiv B_M - (r+1)H.$$

Proof. This is an immediate consequence of the definition of K_X, the formula for $K_{\mathbb{P}^r}$ and the following lemma (cf. §2B for the definition of $f^{-1}(D)$, D a divisor):

(6.20) **Lemma.** *Let* $f: X^r \to Y^r$ *be a regular dominating map of smooth r-dimensional varieties with branch locus B. Then for all rational r-forms ω on Y,*

$$(f^*\omega) = B + f^{-1}((\omega)).$$

Proof. Take any point $x \in X$, let $y = f(x)$ and let $\{x_1, \ldots, x_r\}$ and $\{y_1, \ldots, y_r\}$ be local algebraic coordinates at x and y. Then if

$$\omega = a \cdot dy_1 \wedge \ldots \wedge dy_r$$

and

$$J = \det(\partial(f^*y_i)/\partial x_j),$$

it follows that

$$f^*\omega = f^* a \cdot J \cdot dx_1 \wedge \ldots \wedge dx_r,$$

hence near x:

$$\begin{aligned}
(f^*\omega) &= (J) + (f^*a)\\
&= (J) + f^{-1}((a))\\
&= B + f^{-1}((\omega)). \quad \text{QED}
\end{aligned}$$

§6C. Hilbert Polynomials, Relations with Degree

We turn now to the problem mentioned in §6A: saying something about the dimension of $\mathcal{L}(D)$, the vector space of functions with poles at most D. One of the basic tools for attacking this problem is the beautiful theorem of Hilbert:

(6.21) **Theorem.** *Let M be a finitely generated graded module over the ring* $\mathbb{C}[X_0, \ldots, X_n]$: *i.e., we are given a direct sum decomposition:*

$$M = \bigoplus_{k=k_0}^{\infty} M_k$$

such that for all f homogeneous of degree d, $f \cdot M_k \subset M_{k+d}$. Then there is a polynomial $P_M(t)$ of degree at most n with rational coefficients such that:

$$\dim_{\mathbb{C}} M_k = P_M(k) \text{ for all sufficiently large } k.$$

In particular, if $\mathfrak{A}$ is any homogeneous ideal, the theorem applies to $M = \mathbb{C}[X_0, \ldots, X_n]/\mathfrak{A}$. More particularly, if X is a projective variety, the theorem states that there is a polynomial $P_X(t)$ such that in the notation of (6.6):

$$\dim\begin{bmatrix} \text{degree } k \text{ piece of} \\ \mathbb{C}[X_0, \ldots, X_n]/I(X) \end{bmatrix} = P_X(k), \quad k \gg 0.$$

P_X is called the *Hilbert polynomial* of X and is an extremely important invariant of X. When X is smooth, we can rewrite this formula with linear systems as:

$$\dim L_X(k) = P_X(k) - 1, \quad k \gg 0$$

and since $L_X(k)$ is also complete for large k:

$$\dim |kH| = P_X(k) - 1, \quad k \gg 0$$

where H is a hyperplane section of X.

Proof of (6.21). We use induction on n, starting at $n = -1$, i.e., $\mathbb{C}[X_0,\ldots,X_n] = \mathbb{C}$. Then $M = $ finite-dimensional $\mathbb{C}$-vector space so $M_k = (0), k \gg 0$, so $P(t) \equiv 0$ has the required properties. In the general case, introduce 2 new modules:

$$N' = \{m \in M \mid X_n m = 0\}$$

$$N'' = M/X_n \cdot M.$$

Then N' and N'' are both finitely generated $\mathbb{C}[X_0,\ldots,X_n]$-modules in which multiplication by X_n is 0, hence they are finitely generated $\mathbb{C}[X_0,\ldots,X_{n-1}]$-modules. But for all k, we get an exact sequence of vector spaces:

$$0 \longrightarrow N'_k \longrightarrow M_k \xrightarrow{\text{mult. by } X_n} M_{k+1} \longrightarrow N''_{k+1} \longrightarrow 0.$$

By the induction hypothesis, if $k \gg 0$,

$$\dim N'_k = P'(k)$$

$$\dim N''_k = P''(k)$$

are polynomials of degree at most $n - 1$, hence

$$\dim M_{k+1} - \dim M_k = \dim N''_{k+1} - \dim N'_k$$
$$= P''(k+1) - P'(k).$$

But for any polynomial $f(t)$ of degree d over $\mathbb{Q}$, there is a polynomial $g(t)$ of degree $d+1$ over $\mathbb{Q}$ such that

$$g(t+1) - g(t) \equiv f(t).$$

[Just note that $(t+1)^d - t^d = dt^{d-1} + $ lower order terms, and use induction on degree]. So we may find a polynomial $Q(k)$ of degree n such that

$$Q(k+1) - Q(k) \equiv P''(k+1) - P'(k).$$

Therefore

$$\dim M_{k+1} - Q(k+1) = \dim M_k - Q(k), \quad \text{if} \quad k \gg 0,$$

hence

$$\dim M_k = Q(k) + \text{cnst.} \quad \text{if} \quad k \gg 0. \quad \text{QED}$$

Let's look at a few examples:

i) $M = \mathbb{C}[X_0,\ldots,X_n]$ itself. Then

$$(6.22) \quad \dim M_k = \dim \left\{ \begin{array}{c} \text{vector sp. of polyn.} \\ f \text{ of degree } k \end{array} \right\} = \frac{(k+n)\ldots(k+1)}{n!}, \text{ all } k \geq 0.$$

$$= \frac{k^n}{n!} + \text{lower order terms.}$$

[In fact, for $n = 0$, $M_k = \mathbb{C} \cdot X_0^k$, so dim $M_k = 1$, all $k \geq 0$. In general you can check this by induction on n just as in the proof of (6.21).]

ii) $M = \mathbb{C}[X_0, \ldots, X_n]/(f)$, f homogeneous of degree d. Then we get an exact sequence:

$$0 \longrightarrow \mathbb{C}[X]_{k-d} \xrightarrow{\text{mult. by } f} \mathbb{C}[X]_k \longrightarrow \mathbb{C}[X]/(f)_k \longrightarrow 0,$$

hence

(6.23)
$$\dim \mathbb{C}[X]/(f)_k = \binom{k+n}{n} - \binom{k+n-d}{n}$$

$$= d \, \frac{k^{n-1}}{(n-1)!} + \text{lower oder terms}$$

iii) $M = \mathbb{C}[X_0, \ldots, X_n]/I(S)$, where $S = \{x_1, \ldots, x_s\} \subset \mathbb{P}^n$ is a finite set. Choosing suitable coordinates, we can assume that $x_i \notin V(X_0)$ for any i. Then for all k we get an exact sequence:

$$0 \longrightarrow I(S)_k \longrightarrow \mathbb{C}[X]_k \xrightarrow{\phi} \sum_{i=1}^{s} \mathbb{C}$$

$$\phi(f) = \left(\frac{f}{X_0^k}(x_1), \ldots, \frac{f}{X_0^k}(x_s) \right)$$

But if $k \geq s$ for instance, one can find polynomials f_i of degree k such that

$$f_i(x_j) = 0, \qquad i \neq j$$
$$f_i(x_i) \neq 0$$

hence ϕ is surjective for $k \geq s$. Therefore

(6.24) $\dim(\mathbb{C}[X]/I(S))_k = s, \quad \text{if} \quad k \geq s.$

As these examples suggest, it is more natural to expand Hilbert polynomials $P(t)$ as a sum of binomial coefficients:

(*)
$$a_n \binom{t}{n} + a_{n-1} \binom{t}{n-1} + \ldots\ldots + a_0,$$

rather than as a sum of powers:

$$a_n t^n + a_{n-1} t^{n-1} + \ldots\ldots + a_0.$$

The reason for that is the simple fact that if P is a rational polynomial of degree n such that $P(k) \in \mathbb{Z}$ for any set of $n+1$ consecutive values, then when P is written out in the form (*) the coefficients a_i are integers and conversely if the $a_i \in \mathbb{Z}$, then $P(k) \in \mathbb{Z}$ for all k. In particular, the leading coefficient of P is always of the form $a(t^n/n!)$, with $a \in \mathbb{Z}$. For the Hilbert polynomial, we can interpret this coefficient as follows:

(6.25) **Theorem.** *Let $X^r \subset \mathbb{P}^n$ be a projective variety of degree d. Then its Hilbert polynomial has the form:*

$$P_X(t) = d \cdot \frac{t^r}{r!} + \text{(lower order terms)}.$$

First Proof. In example (ii) above, we saw that this is true for hypersurfaces. We can reduce the general case to that of a hypersurface by projection: choose $L^{n-r-2} \subset \mathbb{P}^n$ so that $L \cap X = \phi$ and so that $p_l : X \to X' \subset \mathbb{P}^{r+1}$ is birational (possible by (3.19)). Then $\deg X' = \deg X$ by (5.5) so it remains to show that the Hilbert polynomials P_X and $P_{X'}$ differ by terms of degree $< r$. In suitable coordinates, $L = V(X_0, \ldots, X_{r+1})$, and the homogeneous coordinate rings of X and X' look like:

$$R = \mathbb{C}[X_0, \ldots, X_n]/I(X)$$
$$R' = \mathbb{C}[X_0, \ldots, X_{r+1}]/(f) \qquad (X' = V(f)).$$

Since the quotient field of R, resp. R', is just $\mathbb{C}(X)(X_0)$, resp. $\mathbb{C}(X')(X_0)$, the fact that $p_L : X \to X'$ is birational implies that R and R' have the same quotient field. Moreover as we noted in (2.28), when you project a variety X from a center disjoint from X, the homogeneous coordinate ring of X is a finite module over that of its image. Thus we have a very well known algebraic situation:

$$\left| \begin{array}{l} R \supset R' \qquad \text{2 domains,} \\ R \text{ a finitely generated } R'\text{-module,} \\ R, R' \text{ same quotient field.} \end{array} \right.$$

In such a case R is not "much bigger" than R', namely for some $\delta \in R'$,

(*) $$\frac{1}{\delta} R' \supset R.$$

In fact, if $x_1, \ldots, x_t \in R$ generate R as R'-module, then each x_i can be written a_i/b_i, $a_i, b_i \in R'$. Then let $\delta = \prod b_i$. (*) also says that $\delta \cdot R \subset R'$. Looking at it this way, one sees that when R and R' are graded and $\delta = \sum_k \delta_k$, then for each k, we must have $\delta_k \cdot R \subset R'$ too. So we may as well assume our δ is homogeneous, say of degree σ. Now to compare Hilbert polynomials, just consider the vector spaces:

$$\delta \cdot R_{k-\sigma} \subset R'_k \subset R_k.$$

Therefore

$$P_X(k - \sigma) \leqq P_{X'}(k) \leqq P_X(k), \quad \text{if} \quad k \gg 0. \qquad \text{QED}$$

Second Proof. Another approach to this theorem is to show first that for almost all hyperplanes H, H meets X transversely in some Y, where Y is irreducible if $\dim X > 1$ and where, for all $x \in Y$, the local equation $h_x \in \mathcal{O}_{x, \mathbb{P}^n}$ of H generates $I(Y)$ in $\mathcal{O}_{x, X}$: call these "good" hyperplanes.

One then shows that for good hyperplane sections Y:

(6.26) $$I(Y)_d = I(X)_d + h \cdot \mathbb{C}[X]_{d-1}, \quad \text{if} \quad d \gg d_0,$$

$$(h = \text{linear equation of } H \text{ in } \mathbb{C}[X_0, \ldots, X_n], d_0 \text{ as in (6.11)})$$

hence
$$P_Y(t) = P_X(t) - P_X(t - 1),$$
and

$$\text{coeff. of } t^{r-1} \text{ in } P_Y(t) = \text{coeff. of } t^r \text{ in } P_X(t).$$

The theorem then follows by induction on r, since for $r = 0$ it is clear by (6.24).

We sketch the proof of these steps without however going into details which involve considerable commutative algebra. To show there are good hyperplane sections, one uses (4.18) to show that for at least one H, $H \cap X$ is transversal and irreducible. Next, one checks that in fact if $\xi = (\xi_0, \ldots, \xi_n)$ and $H_\xi = V(\Sigma \xi_i X_i)$, then the set of all ξ such that $X \cap H_\xi$ is transversal and irreducible is Zariski-open. This is an exercise in applying the ideas of Chapter 3 to the projection of $\mathcal{H} = \{(x,\zeta) | x \in X, \sum x_i \xi_i = 0\}$
$$\underset{\text{def}}{}$$
onto $\mathbb{C}_\xi^{n+1}$. To get H with the last property, cover X by affines $X - X \cap V(X_i)$ and let R_i be their affine coordinate rings. Let R_i' be the integral closure of R_i in $\mathbb{C}(X)$ and consider the primary decomposition of the submodule R_i of the module R_i', i.e.,

$$R_i = M_{i1} \cap \ldots \cap M_{it}$$

where for some prime ideal $\mathfrak{P}_{ij} \subset R_i$, M_{ij} is $\mathfrak{P}_{ij}$-primary, or

a) $\mathfrak{P}_{ij}^N \cdot R_i' \subset M_{ij}$ and

b) $a \in R_i - \mathfrak{P}_{ij}, m \in R_i', a \cdot m \in M_{ij} \Rightarrow m \in M_{ij}$

(cf. $Z - S$, p. 252). The important point is that if $a \in R_i - \cup_j \mathfrak{P}_{ij}$, then for all $m \in R_i, a \cdot m \in R_i \rightarrow m \in R_i$. Let $Z_{ij} \subset X$ be the subvariety such that $Z_{ij} \not\subset V(X_i)$ and $I_{R_i}(Z_{ij}) = \mathfrak{P}_{ij}$. Then to get a good H, we need only take an H with $H \cap X$ transversal irreducible and such that also $H \not\supset Z_{ij}$ for any i,j. Namely, if $h_i = h/X_i \in R_i$ is the local equation of H in R_i, then $h_i \notin \cup_j \mathfrak{P}_{ij}$.

Since $H \cap X$ is transversal, there is a point $x_i \in H \cap X - V(X_i)$ such that X is smooth at x_i and $dh_i \neq 0$ in $T_{x_i, X}$. Then $h_i \cdot \mathcal{O}_{x_i, X}$ is a prime ideal, hence equal to $I(H \cap X) \cdot \mathcal{O}_{x_i, X}$; therefore $h_i R_i$ and $I_{R_i}(H \cap X)$ differ only by a component that is trivial in $\mathcal{O}_{x_i, X}$. This must be an embedded component since $\sqrt{h_i R_i} = I_{R_i}(H \cap X)$ by the Nullstellensatz. Thus

$$h_i R_i = I_{R_i}(H \cap X) \cap (\text{embedded components}).$$

But $h_i R_i'$ has no embedded components ($Z - S$, vol. I, p. 277), hence $h_i R_i' = I_{R_i}(H \cap X) \cdot R_i'$. Thus if $a \in I_{R_i}(H \cap X)$, $a = bh_i$ for some $b \in R_i'$. And $h_i b \in R_i, h_i \notin \cup_j \mathfrak{P}_{ij}$ implies $b \in R_i$. So $a \in h_i R_i$. This proves that $h_i R_i$ is prime for all i, hence H is good. Finally if $R_X = \mathbb{C}[X_0, \ldots, X_n]/I(X)$ and $f \in R_X$ is homogeneous of degree d and zero on $X \cap H$, it follows that for all $i f/X_i^d \in h_i R_i$, or clearing denominators and writing this homogeneously, that

$$X_i^N f \in h \cdot R_X, \text{ all } i, \text{ some large } N.$$

Therefore by (6.11), if $d \geq d_0, f/h \in R_X$ or $f \in h \cdot R_X$, which proves (6.26).

(6.27) **Corollary.** *If $X^r \subset \mathbb{P}^n$ is a projective variety of degree d and M is a graded $\mathbb{C}[X_0, \ldots, X_n]/I(X)$-module, then*

$$P_M(t) = kd \frac{t^r}{r!} + (\text{lower order terms})$$

where if $K = \text{fraction field of } \mathbb{C}[X_0, \ldots, X_n]/I(X)$ then:
$$k = \dim_K(M \otimes K).$$

Proof. First note that when M is a torsion $\mathbb{C}[X]/I(X)$-module, then $\deg P_M < r$. In fact, if $f \in \mathbb{C}[X_0, \ldots, X_n]$ has degree d', $f \notin I(X)$, and $f \cdot M = (0)$, and if M has l generators $m_1, \ldots, m_l$, of degrees $d_1, \ldots, d_l$, then we get a surjection

$$\bigoplus_{i=1}^{l} [\mathbb{C}[X_0,\ldots,X_n]/(f) + I(X)]_{n-d_i} \longrightarrow M_n$$

$$(a_1,\ldots,a_l) \longrightarrow \sum a_i m_i$$

hence

$$P_M(n) \leq \sum_{i=1}^{l} P_X(n-d_i) - P_X(n-d_i-d')$$

and the right hand side has degree $r-1$. Second, for general M, note that if $m_1,\ldots,m_k \in M$, of degrees $d_1,\ldots,d_k$, are a K-basis of $M \oplus K$, then we get an injection

$$\bigoplus_{i=1}^{k} [\mathbb{C}[X_0,\ldots,X_n]/I(X)]_{n-d_i} \longrightarrow M_n$$

$$(a_1,\ldots,a_k) \longrightarrow \sum a_i m_i$$

with cokernel the *n*th graded piece of a torsion module: hence

$$P_M(n) = \sum_{i=1}^{k} P_X(n-d_i) + (\text{degree} < r \text{ terms}) = kd\,\frac{n^r}{r!} + (\text{lower order terms}). \quad \text{QED}$$

At the other end of the scale, the constant term of P_X plays an extremely important role in the geometry of X. For historical reasons, one usually deals not with $P_X(0)$ but with:

(6.28)
$$p_a(X) \underset{\text{def}}{=} (-1)^r(P_X(0) - 1), \quad r = \dim X$$

which is called the *arithmetic genus* of X. In terms of this invariant, and assuming the existence of good hyperplane sections $X \cap H$ of X (cf. second proof of (6.25)), it follows immediately by induction that:

$$P_X(t) = \frac{(t+r-1)\ldots t}{r!} \deg X + \frac{(t+r-2)\ldots t}{(r-1)!}[1 - p_a(X \cap H_1 \cap \ldots \cap H_{r-1})]$$

$$+ \ldots + t\cdot[1 + (-1)^{r-1}p_a(X \cap H_1)] + [1 + (-1)^r p_a(X)],$$

where for all i, $X \cap H_1 \cap \ldots \cap H_{i+1}$ is a good hyperplane section of $X \cap H_1 \cap \ldots \cap H_i$. In other words, the whole Hilbert polynomial is determined by the degree of X and the arithmetic genera of the general intersections $X \cap L$, (L linear) of dimensions $1, 2, \ldots, r$.

One reason for the importance of the arithmetic genus is its striking invariance properties: if X_1, X_2 are 2 varieties and $Z: X_1 \to X_2$ is a biregular correspondence between them, then $p_a(X_1) = p_a(X_2)$; moreover, if X_1 and X_2 are smooth and $Z: X_1 \to X_2$ is any birational correspondence between them, then $p_a(X_1) = p_a(X_2)$ just as with the geometric genus. However, the birational invariance of the arithmetic genus is very deep. 2 proofs are known, one depending on the Hodge theory of harmonic differentials on Kähler manifolds and the other on factorization of birational correspondences via resolution of singularities of their fundamental sets.

Appendix to Chapter 6. The Weil-Samuel Algebraic Theory of Multiplicity

By combining the analytic techniques of Chapter 4 with a varient of the Hilbert polynomial, called the Hilbert-Samuel polynomial, we can give a purely algebraic definition of $\mathrm{mult}(\varphi)$ and open the way to the deeper study of its properties. However, we will need to assume more standard facts both from commutative algebra and complex analysis (such as the quoted results (4.3) in Chapter 4) to do this satisfactorily, and hence we have separated this section from the body of the text. The results proved here will not be used later. To some extent, they are more technical and specialized than the rest of this book. However, those results are also one of the central themes of the 1935–1955 period in which algebraic geometry sought stronger algebraic tools to give precision to its earlier intuitions (and to extend them to characteristic p). The goal was to make rigorous the classical intersection or Schubert calculus (cf. Samuel [1] or Kleiman [1]).

We begin by considering the Nullstellensatz correspondence in the analytic case and relating it to the correspondence in the algebraic case. Consider the following sets:

(A.1)
$$\left\{ \begin{array}{l} \text{ideals } \mathfrak{A} \subset \mathbb{C}[X_1,\ldots,X_n]_{(X)} \\ \text{with } \mathfrak{A} = \sqrt{\mathfrak{A}} \end{array} \right\} \quad \overset{V}{\underset{I}{\longleftrightarrow}} \quad \left\{ \begin{array}{l} \text{closed algebraic} \\ \text{subsets } X \subset \mathbb{C}^n \\ \text{all of whose components} \\ \text{contain } 0 \end{array} \right\}$$

$$\left\{ \begin{array}{l} \text{ideals } \mathfrak{A} \subset \mathbb{C}\{X_1,\ldots,X_n\} \\ \text{with } \mathfrak{A} = \sqrt{\mathfrak{A}} \end{array} \right\} \quad \overset{V}{\underset{I}{\longleftrightarrow}} \quad \left\{ \begin{array}{l} \text{germs of analytic subsets} \\ X_0 \subset U \subset \mathbb{C}^n \text{ at } (0,\ldots,0) \end{array} \right\}$$

The usual Nullstellensatz says that the V, I on the top line are inverse bijections and that for arbitrary $\mathfrak{A}, I(V(\mathfrak{A})) = \sqrt{\mathfrak{A}}$. The analytic Nullstellensatz says that the V, I on the next line are also inverse bijections and that for arbitrary $\mathfrak{A}, I(V(\mathfrak{A})) = \sqrt{\mathfrak{A}}$: cf. Gunning-Rossi, p. 97. But there are also natural vertical arrows assigning to an algebraic $X \subset \mathbb{C}^n$ the germ X_0^{an} of analytic sets that it defines at 0; or assigning to the germ $X_0 \subset U \subset \mathbb{C}^n$ its Zariski-closure $\overline{X}_0 \subset \mathbb{C}^n$ (i.e., the smallest closed algebraic set containing the germ X). Clearly, starting from an algebraic $X \subset \mathbb{C}^n$ all of whose components contain 0, forming first X_0^{an}, then $(\overline{X_0^{an}})$, we find:

$$(\overline{X_0^{an}}) = X.$$

Corresponding to these vertical maps, we get vertical maps on the ideal side too. These will be:

$$\text{given } \mathfrak{A} \subset \mathbb{C}[X_1,\ldots,X_n]_{(X)}, \text{ from } \mathfrak{A} \cdot \mathbb{C}\{X_1,\ldots,X_n\}$$

$$\text{given } \mathfrak{A} \subset \mathbb{C}\{X_1,\ldots,X_n\}, \text{ from } \mathfrak{A} \cap \mathbb{C}[X_1,\ldots,X_n]_{(X)}.$$

Here we have 2 facts that make the situation reasonably simple:

(A.2) **Proposition.** a) (Chevalley) if $\mathfrak{A} \subset \mathbb{C}[X_1,\ldots,X_n]_{(X)}$ satisfies $\mathfrak{A} = \sqrt{\mathfrak{A}}$, then

$$\mathfrak{A}' = \mathfrak{A} \cdot \mathbb{C}\{X_1,\ldots,X_n\} \text{ satisfies } \mathfrak{A}' = \sqrt{\mathfrak{A}'}$$

   b) (Krull) *for all* $\mathfrak{A} \subset \mathbb{C}[X_1,\ldots,X_n]_{(X)}$,

$$\mathbb{C}[X_1,\dots,X_n]_{(X)} \cap \mathfrak{A} \cdot \mathbb{C}\{X_1,\dots,X_n\} = \mathfrak{A}.$$

Here (b) is a consequence of the same result of Krull that we have used before. Namely, his result is that

$$R \cap \mathfrak{A} \cdot \hat{R} = \mathfrak{A}$$

and since it is clear that

$$\mathfrak{A} \subseteq \mathbb{C}[X]_{(X)} \cap \mathfrak{A} \cdot \mathbb{C}\{X\} \subseteq \mathbb{C}[X]_{(X)} \cap \mathfrak{A} \cdot \mathbb{C}[[X]],$$

applying his result to $R = \mathbb{C}[X]_{(X)}$, we get (b). Applying Krull's result to $R = \mathbb{C}\{X\}$, we find in fact that we have inclusions:

$$\mathbb{C}[X]_{(X)}/\mathfrak{A} \subset \mathbb{C}\{X\}/\mathfrak{A} \cdot \mathbb{C}\{X\} \subset \mathbb{C}[[X]]/\mathfrak{A} \cdot \mathbb{C}[[X]].$$

Now $\mathbb{C}[[X]]/\mathfrak{A} \cdot \mathbb{C}[[X]]$ is just the formal completion of the ring $\mathbb{C}[X]_{(X)}/\mathfrak{A}$. Chevalley proved (see Z-S, vol. II, p. 320) that when $\mathbb{C}[X]_{(X)}/\mathfrak{A}$ had no nilpotent elements, i.e., $\mathfrak{A} = \sqrt{\mathfrak{A}}$, then its formal completion has the same property. It follows that $\mathbb{C}\{X\}/\mathfrak{A} \cdot \mathbb{C}\{X\}$ has no nilpotents too, which gives us (a).

Combining this with the analytic Nullstellensatz, we get:

(A.3) **Corollary.** *If $X \subset \mathbb{C}^n$ is a closed algebraic set containing 0, and $f_1,\dots,f_m$ generate its affine ideal, then $f_1,\dots,f_m$ also generate the ideal of analytic functions in $\mathbb{C}\{X_1,\dots,X_n\}$ zero on the germ at 0 defined by X. Hence*

$$\mathcal{O}^{an}_{X,0} \cong \mathbb{C}\{X_1,\dots,X_n\}/(f_1,\dots,f_m).$$

Moreover, we may use the analytic Nullstellensatz plus ideal decomposition in the noetherian ring $\mathbb{C}\{X_1,\dots,X_n\}$ to deduce:

(A.4) **Corollary.** *Every germ of analytic space $X \subset U \subset \mathbb{C}^n$ at 0 has a unique decomposition:*

$$X \cap U' = X_1 \cup \dots \cup X_s$$

where $U' \subset U$ is a smaller neighborhood of 0, $X_i \subset U'$ are analytic sets and where the X_i are irreducible (i.e., for all $U'' \subset U$, if $X_i \cap U'' = Y \cup Z$, then in some smaller $U''' \subset U''$, $X_i \cap U''' = Y \cap U'''$ or $= Z \cap U'''$). Moreover X is an irreducible germ if and only if $I(X) \subset \mathbb{C}\{X_1,\dots,X_n\}$ is prime.

In fact, suppose we define X to be topologically unibranch at 0 if there is a fundamental system of neighborhoods U_n of 0 such that

$$U_n \cap X - U_n \cap \text{Sing } X$$

are connected. Then following the ideas of Chapter 4, we get a local version of (4.16):

(A.5) **Proposition.** *A germ of analytic space $X \subset U \subset \mathbb{C}^n$ at 0 is topologically unibranch if and only if it is irreducible.*

Proof. In fact if $X \cap U'$ breaks up as $X_1 \cup X_2$ while X is topologically unibranch, then shrinking U' if necessary we may assume $X \cap U' - \text{Sing } X \cap U'$ is a connected complex manifold. Such an analytic space is certainly irreducible, so $X_i \supset X \cap U' - \operatorname{Sing} X \cap U'$ for some i. Then $X_i \supset \overline{X \cap U' - \operatorname{Sing} X \cap U'} = X \cap U'$. Conversely, if X is not topologically unibranch, then for all sufficiently small U', $X \cap U' - \operatorname{Sing} X \cap U'$ is disconnected: let Y_α be its components. By (4.7), their closures $\overline{Y}_\alpha$ in U' are analytic sets, so we have decomposed $X \cap U'$ as $\cup Y_\alpha$. By (4.3)(i), this is a locally finite union, hence

$$Z = \bigcup_{\substack{(\text{those } \alpha \text{ for} \\ \text{which } 0 \notin Y_\alpha)}} \overline{Y}_\alpha$$

is closed. Replacing U' by $U' - Z$, we may assume that $0 \in \overline{Y}_\alpha$, all α. By assumption there is more than one component and as $\overline{Y}_\alpha \cap Y_\beta = \phi$, no $\overline{Y}_\alpha$ contains a whole neighborhood of 0 in X. Thus the germ of X at 0 is reducible.

(A.6) Definition. *When $\mathcal{O}^{an}_{x,X}$ is a domain, let $K^{an}_{x,X}$ be its field of fractions.*

Another basic fact relating the algebra of the rings $\mathcal{O}^{an}_{x,X}$ to the geometry is:

(A.7) Proposition. *Let $\varphi \colon X \longrightarrow Y$ be a holomorphic map of analytic sets. Let $x \in X$ satisfy:*

$$\{x\} = a \text{ component of } \varphi^{-1} \varphi(x).$$

Then if $y = \varphi(x)$, $\mathcal{O}^{an}_{x,X}$ is a finitely generated $\mathcal{O}^{an}_{y,Y}$ – module.

Proof. It is easy to see that if $Y \subset V \subset \mathbb{C}^m$, then X is isomorphic to $X' = \{x, \varphi(x) | x \in X\} \subset U \times V \subset \mathbb{C}^{n+m}$. Replacing X by X', we may assume φ is the projection $p \colon \mathbb{C}^{n+m} \longrightarrow \mathbb{C}^n$. Factoring the projection into 1-dimensional pieces, we are reduced to the following: if $p \colon \mathbb{C}^{n+1} \longrightarrow \mathbb{C}^n$ is a projection and $X \subset U \subset \mathbb{C}^{n+1}$ is an analytic set such that $X \cap p^{-1} 0 = \{0\}$, then $\mathcal{O}^{an}_{0,X}$ is a finite generated $\mathbb{C}\{X_1, \ldots, X_n\}$ – module. In this case, at least one of the equations f defining X must satisfy $f(0, \ldots, 0, X_{n+1}) \not\equiv 0$. Therefore by the Weierstrass Preparation Theorem, we may assume that

$$f = X^d_{n+1} + a_1(X_1, \ldots, X_n) X^{d-1}_{n+1} + \ldots + a_d(X_1, \ldots, X_n).$$

It follows from the 2nd half of the preparation theorem that $X^{d-1}_{n+1}, \ldots, X_{n+1}, 1$ are a basis of $\mathcal{O}^{an}_{0,X}$ as a $\mathbb{C}\{X_1, \ldots, X_n\}$ module.   QED

Now suppose we look at the situation of Chapter 3 in which mult has been defined:

$$\varphi \colon X^r \longrightarrow Y^r$$

a regular map of affine varieties, $x \in X, y = \varphi(x)$ satisfying

    i) $\{x\} = a$ component of $\varphi^{-1}(y)$,
    ii) Y topologically unibranch at y.
Then we claim that:

(A.8) Formula of Weil.

$$\operatorname{mult}_x \varphi = \dim_{K^{an}_{y,Y}} \left[\mathcal{O}^{an}_{x,X} \otimes_{\mathcal{O}^{an}_{y,Y}} K^{an}_{y,Y} \right].$$

More generally, this formula makes sense, and is true, if $X \subset U \subset \mathbb{C}^n, Y \subset V \subset \mathbb{C}^n$

are any 2 analytic sets, φ is analytic and $x \in X, y = \varphi(x)$ satisfy (i), (ii). This is because, exactly as in Chapter 3, there will be $U' \subset U, V' \subset V$ such that res φ is a proper map from $X \cap U'$ to $Y \cap V'$ which is an unramified covering over an open, dense connected subset Y_0' of $Y \cap V'$. We define $\text{mult}_x \varphi$ to be the degree of this covering. Here is a sketch of the proof of (A.8).

Step I. Reduction to the case X topologically unibranch at x too.

Proof. In $\mathcal{O}_{x,X}^{an}$, the ideal (0) can be written as an irredundant intersection of prime ideals:

$$(0) = \mathfrak{P}_1 \cap \ldots \cap \mathfrak{P}_s, \mathfrak{P}_i \nsubseteq \mathfrak{P}_j.$$

Generators of $\mathfrak{P}_i$ define a germ of analytic set $X_i \subset X$ near x, and near x, $X = X_1 \cup \ldots \cup X_s$, where

$$\mathcal{O}_{x,X_i}^{an} \cong \mathcal{O}_{x,X}^{An}/\mathfrak{P}_i,$$

so X_i is topologically unibranch at x by (A.5). Then on the one hand, the covering $\varphi^{-1}(Y_0') \longrightarrow Y_0'$ breaks up into the disjoint union of coverings

$$X_i \cap \varphi^{-1}(Y_0') \longrightarrow Y_0'$$

hence if $\varphi_i =$ restriction of φ to X_i,

$$\text{mult}_x \varphi = \sum_{i=1}^s \text{mult}_x \varphi_i.$$

And on the other hand, writing $K = K_{y,Y}^{an}, R = \mathcal{O}_{x,X}^{an} \otimes_{\mathcal{O}_{y,Y}^{an}} K_{y,Y}^{an}$, then *in R*:

$$(0) = \mathfrak{P}_1' \cap \ldots \cap \mathfrak{P}_s'$$

$$\mathfrak{P}_i' = \mathfrak{P}_i \cdot R.$$

R is a finite-dimensional commutative K-algebra without nilpotents, so this means*

$$R = \bigoplus_{i=1}^s R/\mathfrak{P}_i',$$

i.e.,

$$\dim_K[\mathcal{O}_{x,X}^{an} \otimes K] = \sum_{i=1}^s \dim_K[\mathcal{O}_{x,X_i}^{an} \otimes K].$$

Step II. As in the proof of (2.7), we reduce to the case where $X \subset U \subset \mathbb{C}^{n+1}$, $Y \subset V \subset \mathbb{C}^n$, $\varphi = p$, a projection, $x = y = 0$.

Step III. In this case, X_{n+1} generates $\mathcal{O}_{0,X}^{an}$ as $\mathcal{O}_{0,Y}^{an}$ – algebra, hence it generates the field $K_{0,X}^{an}$ over $K_{0,Y}^{an}$. Let d be the degree of this field extension. Then on X, X_{n+1} satisfies an equation

$$g \equiv a_0(X_1, \ldots, X_n) X_{n+1}^d + \ldots + a_d(X^1, \ldots, X_n) = 0.$$

---

*This is OK because R is a localization of $\mathcal{O}_{x,X}^{an}$ and the $\mathfrak{P}_i$ are distinct, so all the $\mathfrak{P}_i'$ which are not R are still distinct.

Every other defining equation f_i of X must satisfy:

$$b_i(X_1,\ldots,X_n)\cdot f_i(X_1,\ldots,X_{n+1}) = h_i(X_1,\ldots,X_{n+1})\cdot g(X_1,\ldots,X_{n+1}).$$

If $\Delta = $ discriminant of g and $T \subset Y$ is defined by $a_0\cdot \Pi b_i\cdot \Delta = 0$, then over $Y - T$, X is defined by the single equation g and is an unramified d-sheeted covering. Therefore $d = \text{mult}_x\,\varphi$.   QED

This is the first, partially algebraic, formula for mult. The next step is to define the Hilbert-Samuel polynomial.

(A.9) Theorem-Defintion. *Let $R = \mathcal{O}_{x,X}$, where X is a variety of dimension n, $\mathfrak{M}$ its maximal ideal, $\mathfrak{A} \subset R$ any ideal such that $\mathfrak{A} \supset \mathfrak{M}^k$ for some k. Let M be a finitely generated R-module. Then $M/\mathfrak{A}^l\cdot M$ is a finitely generated $R/\mathfrak{M}^{kl}$-module, hence it has finite dimension as complex vector space. Then there is a polynomial $P(t)$ of degree at most n called the Hilbert-Samuel polynomial such that:*

$$\dim_{\mathbb{C}}(M/\mathfrak{A}^l\cdot M) = P(l), \text{ if } l \gg 0.$$

If

$$P(t) = e\cdot\frac{t^n}{n!} + \text{(lower order terms)},$$

then $e = e_R(\mathfrak{A};M)$ or $e_x(\mathfrak{A};M)$ is called the multiplicity of M with respect to $\mathfrak{A}$. If $M = R, e$ is simply denoted $e_R(\mathfrak{A})$ or $e_x(\mathfrak{A})$.

Proof. Look at the graded group

$$gr\,M = \bigoplus_{k=0}^{\infty} \mathfrak{A}^k\cdot M/\mathfrak{A}^{k+1}\cdot M.$$

Choose $y_1,\ldots,y_t\in\mathfrak{A}$ generating $\mathfrak{A}$ as ideal and make $gr\,M$ into a graded $\mathbb{C}[Y_1,\ldots,Y_t]$-module by letting Y_i act by

$$\mathfrak{A}^k M/\mathfrak{A}^{k+1}M \xrightarrow{\text{mult. by } y_i} \mathfrak{A}^{k+1}\cdot M/\mathfrak{A}^{k+2}\cdot M.$$

Now $M/\mathfrak{A}\cdot M$ is a finite-dimensional complex vector space: let $e_1,\ldots,e_N$ be a basis. Then one sees immediately that the e_i generate $gr\,M$ as $\mathbb{C}[Y_1,\ldots,Y_t]$-module. Therefore by the Hilbert polynomial theorem (6.21), for k large

$$\dim_{\mathbb{C}}(\mathfrak{A}^k M/\mathfrak{A}^{k+1}M)$$

is a polynomial in k of degree at most $t-1$. Therefore, as in the proof of (6.21), for k large, $\dim_{\mathbb{C}}(M/\mathfrak{A}^k M)$ is a polynomial in k of degree at most t.

We want to improve the bound on the degree of P. Since $\dim X = n$, there exist $f_1,\ldots,f_n\in R$ such that $\{x\}$ is a component of $V(f_1,\ldots,f_n)$. Therefore, by the Nullstellensatz,

$$\mathfrak{M} = \sqrt{(f_1,\ldots,f_n)}.$$

Replacing f_i by $f_i^l, l \gg 0$, we may assume $f_i\in\mathfrak{A}$. Let $\mathfrak{B} = (f_1,\ldots,f_n)$. Then $\mathfrak{B}^l \subset \mathfrak{A}^l$, hence

$$P(l) = \dim_{\mathbb{C}}(M/\mathfrak{A}^l\cdot M)$$
$$\leq \dim_{\mathbb{C}}(M/\mathfrak{B}^l M)$$

and by the first part of the proof, $\dim_{\mathbb{C}}(M/\mathfrak{B}^l M)$ is a polynomial of degree at most n. Therefore the same holds for P.   QED

Note that the same result and proof works for more general local rings R: thus if $\mathfrak{M}$ = maximal ideal of R and 1) R is a $\mathbb{C}$ − algebra with $R/\mathfrak{M} = \mathbb{C}$, 2) R has dimension $\leq n$ in the sense that $\mathfrak{M} = \sqrt{(f_1,\ldots,f_n)}$ for some $f_1,\ldots,f_n \in \mathfrak{M}$, then the above proof works verbatim. This applies, e.g. to $\mathcal{O}^{an}_{x,X}$. (Actually the theorem is valid for arbitrary noetherian local rings R if "$\dim_{\mathbb{C}}$" is replaced by "length as R − module" and $\dim R$ is interpreted as just mentioned, and only small changes in the proof are needed).

This definition gives one way to get a purely algebraic definition of $\operatorname{mult}_x \varphi$. The result is this:

(A.10) **Theorem.** *If* $\varphi : X \longrightarrow Y$ *is a regular map of affine varieties and* $x \in X$, $y = \varphi(x)$ *satisfy*

  i) $\{x\}$ = *a component of* $\varphi^{-1}(y)$
  ii) y *smooth on* Y,

then we have the formula of Samuel:

$$\operatorname{mult}_x \varphi = e_x(\mathfrak{M}_{y,Y} \cdot \mathcal{O}_{x,X}).$$

Proof. We want to play on the 4 rings:

$$\mathcal{O}^{an}_{x,X} \xleftarrow{\quad\phi^*\quad} \mathcal{O}^{an}_{y,Y}$$
$$\cup \qquad\qquad\qquad \cup$$
$$\mathcal{O}_{x,X} \xleftarrow{\quad\phi^*\quad} \mathcal{O}_{y,Y}.$$

The idea is that, according to (A.8)

$$\operatorname{mult}_x \varphi = \dim_{K^{an}_{y,Y}} \left[K^{an}_{y,Y} \otimes_{\mathcal{O}^{an}_{x,X}} \mathcal{O}^{an}_{x,X} \right].$$

On the other hand, by (A.3), we see that all the finite-dimensional truncations

$$\mathcal{O}_{x,X}/\mathfrak{M}^N_{x,X}, \quad \mathcal{O}^{an}_{x,X}/(\mathfrak{M}^{an}_{x,X})^N$$

are isomorphic, hence we have

$$e_{\mathcal{O}_{x,X}}(\mathfrak{M}_{y,Y} \cdot \mathcal{O}_{x,X}) = e_{\mathcal{O}^{an}_{x,X}}(\mathfrak{M}^{an}_{y,Y} \cdot \mathcal{O}^{an}_{x,X}).$$

By assumption (a), $\mathcal{O}^{an}_{x,X}$ is a finitely generated $\mathcal{O}^{an}_{y,Y}$-module, hence we have directly, by the definition

$$e_{\mathcal{O}^{an}_{x,X}}(\mathfrak{M}^{an}_{y,Y} \cdot \mathcal{O}^{an}_{x,X}) = e_{\mathcal{O}^{an}_{y,Y}}(\mathfrak{M}^{an}_{y,Y} ; \mathcal{O}^{an}_{x,X}).$$

By assumption (b), $\mathcal{O}^{an}_{y,Y} \cong \mathbb{C}\{X_1,\ldots,X_n\}$, thus the theorem reduces to the calculation:

(A.11) **Lemma.** *Let* $R = \mathbb{C}\{X_1,\ldots,X_n\}$ *and let* M *be a finitely generated* R-*module. Let* K *be the quotient field of* R. *Then*

$$e_R((X_1,\ldots,X_n); M) = \dim_K(M \otimes_R K).$$

The proof is similar to that of (6.27): first check that if M is a torsion R-module, then $gr\, M$ is a torsion $\mathbb{C}[X_1,\dots,X_n]$-module, hence by the results on Hilbert polynomials, $e = 0$. Also if $M = R$, then $X^\alpha, |\alpha| < k$, are a basis of $M/(X_1,\dots,X_n)^l M$, hence

$$P_M(t) = \binom{t+n-1}{n}$$

hence $e = 1$. In the general case, if $k = \dim_K(M \otimes_R K)$, one has exact sequences:

$$R^k \longrightarrow M \longrightarrow T_1 \longrightarrow 0$$

$$M \longrightarrow R^k \longrightarrow T_2 \longrightarrow 0$$

T_i torsion R-modules, hence, if $\mathfrak{A} = (X_1,\dots,X_n)$,

$$(R/\mathfrak{A}^l)^k \longrightarrow M/\mathfrak{A}^l M \longrightarrow T_1/\mathfrak{A}^l T_1 \longrightarrow 0$$

$$M/\mathfrak{A}^l M \longrightarrow (R/\mathfrak{A}^l)^k \longrightarrow T_2/\mathfrak{A}^l T_2 \longrightarrow 0.$$

This shows

$$\left| \dim_{\mathbb{C}}(R/\mathfrak{A}^l)^k - \dim_{\mathbb{C}} M/\mathfrak{A}^l M \right| < \binom{\text{polyn. in } l \text{ of}}{\text{degree } n-1}$$

hence

$$e_R(\mathfrak{A}\,;M) = e_R(\mathfrak{A},R^k) = k. \quad \text{QED}$$

I want to emphasize the rather startling consequence of this formula:

(A.12) **Corollary.** $\text{mult}_x \varphi$ *depends only on the ideal* $\mathfrak{M}_{y,Y} \cdot \mathcal{O}_{x,X}$.

We can give formulae for the other types of multiplicities introduced in Chapter 5 too. For instance, for the multiplicity of a point, we get:

(A.13) **Corollary.** *Let* X^r *be an affine variety,* $x \in X$, *and let*

$$p: X \longrightarrow \mathbb{C}^r$$

be a projection with

$$p(x) = 0$$

$$p^{-1}(0) \cap E^*_{x,X} = \{0\}.$$

Let p be defined by functions $t_1,\dots,t_r$. Then

$$\text{mult}_x(X) = e_x((t_1,\dots,t_r)).$$

Proof. Combine (5.11) with (A.10).

We also gave a definition in Chapter 5 for the intersection multiplicity of 2 subvarieties X^t, Y^{r-t} of a smooth variety Z^r at an isolated point $\{P\}$ of intersection, depending however on the certain auxiliary choices. We now get the beautiful intrinsic formula:

(A.14) **Corollary.** *If $\{P\}$ is a component of $X^t \cap Y^{r-t}, X, Y$ being subvarieties of a smooth variety Z^r, then*

$$i(P; X \cap Y) = e_{(P,P)}(I_\Delta)$$

where $I_{\bar{\Delta}} \subset \mathcal{O}_{(P,P),X \times Y}$ is the ideal generated by the functions $g(x,y) = f(x) - f(y)$, all $f \in \mathcal{O}_{P,Z}$.

Proof. If $t_1,\ldots,t_r \in \mathcal{O}_{P,Z}$ are local analytic coordinates, then by (3.29) on $Z \times Z$ near (P,P) the functions $g_i(x,y) = t_i(x) - t_i(y)$ generate the full ideal of functions vanishing on the diagonal. Therefore I_Δ is generated by $g_1,\ldots,g_r$ and (A.14) follows immediately from (A.10) and the definition of i.

For $\text{mult}_x X$ there is an even nicer formula, namely:

$$(A.15) \qquad \text{mult}_x(X) = e_x(\mathfrak{M}_{x,X}).$$

In fact, Z-S, vol. II, p. 294 prove that if $\dim X = r$, then for any ideal $\mathfrak{A} \subset \mathcal{O}_{x,X}$ with $\mathfrak{A} \supset \mathfrak{M}^N_{x,X}$,

$$e_x(\mathfrak{A}) = e_x((t_1,\ldots,t_r))$$

for any $t_1,\ldots,t_r \in \mathfrak{A}$ which are chosen "sufficiently generally". Applying this with $\mathfrak{A} = \mathfrak{M}_{x,X}$, (A.15) reduces to (A.13). However, there is a very pretty geometric way to prove (A.15) that I would like to sketch. We assume that the given variety $X^r \subset \mathbb{P}^n$ is not a cone with vertex x. The idea is to mirror the formula:

$$\deg X - \text{mult}_x(X) = \deg(\text{res}_X p_x) \cdot \deg p_x(X)$$

in graded modules. Thus, say $(X_0,\ldots,X_n)$ are coordinates in $\mathbb{P}^n$ with $x = (1,0,\ldots,0)$. Consider the graded ring homomorphism

$$\phi : \mathbb{C}[X_0,\ldots,X_n] \longrightarrow \bigoplus_{k=0}^{\infty} \mathcal{O}_{x,\mathbb{P}^n} \cdot X_0^k$$

given by $\phi(f) = (f/X_0^k) \cdot X_0^k$ if f is homogeneous of degree k. It induces a graded homomorphism:

$$\psi : R_X \longrightarrow \bigoplus_{k=0}^{\infty} \mathcal{O}_{x,X} \cdot X_0^k.$$

$$\| $$

$$\mathbb{C}[X_0,\ldots,X_n]/I(X)$$

Define the new graded ring:

$$R^0_X = \psi^{-1}\left[\bigoplus_{k=0}^{\infty} \mathfrak{M}^k_{x,X} \cdot X_0^k \right].$$

Note that on the lth piece, the map

$$R_{X,l} \longrightarrow [\mathcal{O}_{x,X}/\mathfrak{M}^l_{x,X}] \cdot X_0^l$$

is surjective: this is because the monomials

$$1; \frac{X_1}{X_0}, \frac{X_2}{X_0}, \ldots, \frac{X_n}{X_0}; \cdots \cdots \left(\frac{X_1}{X_0}\right)^l, \left(\frac{X_1}{X_0}\right)^{l-1} \cdot \left(\frac{X_2}{X_0}\right), \ldots, \left(\frac{X_n}{X_0}\right)^l$$

span $\mathcal{O}_{x,X}/\mathfrak{M}^l_{x,X}$. Thus:

$$\dim_{\mathbb{C}}(R^0_X)_l = \dim_{\mathbb{C}}(R_{X,L}) - \dim_{\mathbb{C}}(\mathcal{O}_{x,X}/\mathfrak{M}^l_{x,X})$$

$$= (\deg X)\frac{l^n}{n!} - e_x(\mathfrak{M}_{x,x})\cdot\frac{l^n}{n!} + \text{(lower order terms)}.$$

Note that $(R_X^0)_1 = \mathbb{C}\cdot X_1 + \ldots + \mathbb{C}\cdot X_n$, hence

$$R_{p_x(X)} = \mathbb{C}[X_1,\ldots,X_n]/I(p_x(X))$$

is a subring of R_X^0. If we now prove

(A.16) Lemma. R_X^0 is a finitely generated $R_{p_x(X)}$-module which gives us a vector space of dimension $\deg(\mathrm{res}_X p_x)$ over the fraction field of $R_{p_x(X)}$.

then by (6.27), we have

$$\dim\,(R_X^0)_l = \deg(\mathrm{res}_X p_x)\cdot\deg p_x(X)\cdot\frac{l^n}{n!} + \text{(lower order terms)}$$

hence

$$\deg X - e_x(\mathfrak{M}_{x,x}) = \deg(\mathrm{res}_X p_x)\cdot\deg p_x(X).$$

This is exactly our old formula from Chapter 5 with $\mathrm{mult}_x X$ replaced by $e_x(\mathfrak{M}_{x,x})$! So $\mathrm{mult}_x X = e_x(\mathfrak{M}_{x,x})$.

Idea of proof of lemma. Take $f \in R_X^0$, homogeneous of degree l. We want to show $R_{p_x(X)}[f]$ is a finitely generated module over $\mathbb{C}[X_1,\ldots,X_n]$. The idea is to consider the embeddings of $X - \{x\}$ in projective space defined by

  i) monomials in $X_1,\ldots,X_n$ of degree l
  ii) these monomials and also f.

The first gives us the l-tuple of $p_x(X) \subset \mathbb{P}^{n-1}$: call it $p_x(X)' \subset \mathbb{P}^N$. The second gives us some map of $X - \{x\}$ into $\mathbb{P}^{N+1}$ such that projecting from a point $y \in \mathbb{P}^{N+1}$ we get $p_x(X)'$:

$$
\begin{array}{ccc}
X - \{x\} & \xrightarrow{\quad \alpha \quad} & \mathbb{P}^{N+1} - \{y\} \\[4pt]
\downarrow{\scriptstyle p_x} & & \downarrow{\scriptstyle p_y} \\[4pt]
p_x(X) & \xrightarrow[\quad l-\text{tuple} \quad]{} & \mathbb{P}^N
\end{array}
$$

One then shows easily that α extends to a regular map from $B_x(X)$ to $\mathbb{P}^{N+1}$ and that y is not in the image. Then if X' is the image, it follows that $R_{X'} = \mathbb{C}[f,X_1^l,\ldots,X_n^l]/I(X')$ is a finitely generated module over $R_{p_x(X)'} = \mathbb{C}[X_1^l,\ldots,X_n^l]$ $/I(p_x X')$, hence so is $R_{p_x(X)}[f]$ over $\mathbb{C}[X_1,\ldots,X_n]$. To get the second assertion, we need to check that the fraction fields of R_X^0 and of R_X are equal (i.e., $\cong \mathbb{C}(X)(X_0)$). But let

$$f_k(X_1,\ldots,X_n)\cdot X_0^{d-k} + f_{k+1}(X_1,\ldots,X_n)\cdot X_0^{d-k-1} + \ldots$$

be an equation for X involving X_0. Then

$$\frac{f_k(X_1,\ldots,X_n)}{X_0^k} \in \mathfrak{M}_{x,x}^{k+1}, \text{ so } f_k(X_1,\ldots,X_n)X_0 \in (R_X^0)_{k+1}$$

so

$$X_0 = \frac{f_k(X_1,\ldots,X_n) \cdot X_0}{f_k(X_1,\ldots,X_n)} \in (\text{fraction field of } R_X^0). \quad \text{QED}$$

To finish the chapter I would like to add at this point an example of a calculation of multiplicity that illustrates a very important subtlety. In many elementary cases, it turns out that

$$\sum_{k=0}^{\infty} \mathfrak{M}_{y,Y}^k \cdot \mathcal{O}_{x,X}/\mathfrak{M}_{y,Y}^{k+1} \cdot \mathcal{O}_{x,X}$$

is a polynomial ring over $\mathcal{O}_{x,X}/\mathfrak{M}_{y,Y} \cdot \mathcal{O}_{x,Y}$ and, when this happens, by an immediate computation one finds:

$$e_x(\mathfrak{M}_{y,Y} \cdot \mathcal{O}_{x,X}) = \dim_{\mathbb{C}}[\mathcal{O}_{x,X}/\mathfrak{M}_{y,Y} \cdot \mathcal{O}_{x,X}].$$

So it has long been felt that the "reasonable" case was when:

(A.17) $\text{mult}_x(\varphi) = \dim_{\mathbb{C}}[\mathcal{O}_{x,X}/\mathfrak{M}_{y,Y} \cdot \mathcal{O}_{x,X}]$

and that only in pathological cases was this false. In fact, (A.17) has a very nice heuristic interpretation: assume φ is a proper finite-to-one map from an affine X to an affine Y and $\{x\} = \varphi^{-1}(\varphi(x))$ for simplicity. Then for almost all $y \in Y$, φ is smooth over y and if $d = \deg(\varphi)$, $\varphi^{-1}(y)$ consists of d points. Let A_y be the algebra of complex-valued functions on the finite set $\varphi^{-1}(y)$. As one expects, these vector spaces A_y fit together into a *vector bundle* $A^{(0)}$ of rank d over this open subset $Y_0 \subset Y$. One would like, now, to extend $A^{(0)}$ to a vector bundle over the whole of Y, with fibres A_y given by the algebras:

$$A_y = R_X/I(y) \cdot R_X$$

(R_X = affine coordinate ring of X, $I(y)$ = maximal ideal in R_Y of functions zero at y). If φ is not smooth over y, one expects A_y to have nilpotents, but one might hope that the dimension does not change and that

$$\bigcup_y A_y = A, \text{ a vector bundle over } Y.$$

When this happens, (A.17) just asserts the equality of the dimension of the "generic" fibres of A and the fibre over $\varphi(x)$.

The simplest example that I know where (A.17) fails is the following:
a) Writing subscripts in $\mathbb{A}^n$ to denote the variables, define

$$f: \mathbb{A}^2_{(s,t)} \longrightarrow \mathbb{A}^4_{(x_1,x_2,x_3,x_4)}$$

by

$$x_1 = s^4$$

$$x_2 = s^3 t$$

$$x_3 = s t^3$$

$$x_4 = t^5. \quad [\text{Note: } s^2 t^2 \text{ is omitted!}]$$

b) Let $X = \text{Image}\,(f)$: this is a variety with prime ideal $\mathfrak{P}$ generated by

$$x_1 x_4 - x_2 x_3$$
$$x_1^2 x_3 - x_2^3$$
$$x_2 x_4^2 - x_3^3$$
$$x_2^2 x_4 - x_3^2 x_1$$

c) X is a cone over $(0,0,0,0)$ and its degree is easily seen to be 4.

d) Let $Y = \mathbb{A}^2_{(x_1,x_4)}$ and let

$$\varphi : X \longrightarrow Y$$

be projection. In fact φ is finite-to-one, proper and $\varphi^{-1}(0,0) = \{(0,0,0,0)\}$.

e) $\deg \varphi = 4$ since $\mathbb{C}(X)$ is generated by $\sqrt[4]{x_4/x_1}$ over $\mathbb{C}(Y)$. But

$$\mathcal{O}_{0,X}/\mathfrak{M}_{0,Y} \cdot \mathcal{O}_{0,X} \cong R_X/(x_1,x_4) \cdot R_X$$
$$\cong \mathbb{C}[x_1,x_2,x_3,x_4]/\mathfrak{P} + (x_1,x_4)$$
$$\cong \mathbb{C}[x_2,x_3]/(x_2^3,x_2 x_3 . x_3^3)$$

which is 5-dimensional.

Chapter 7. Curves and Their Genus

§7A. Existence and Uniqueness of the Non-Singular Model of Each Function Field of Transcendence Degree 1 (after Albanese)

A field K containing $\mathbb{C}$ and finitely generated over it is called a *function field*. Alternately, these are exactly the fields $\mathbb{C}(X)$, for varieties X. In fact, for any K of transcendence degree n, let $x_1, \ldots, x_n \in K$ be a transcendence base. Then K is a finite algebraic extension of $\mathbb{C}(X_1, \ldots, X_n)$: such an extension can be generated by one element. Let x_{n+1} be such an element. Then in $K, x_1, x_2, \ldots, x_{n+1}$ satisfy an irreducible equation $f(X_1, \ldots, X_{n+1}) = 0$, unique up to scalars. Therefore

$$K = \text{fraction field of } \mathbb{C}[X_1, \ldots, X_{n+1}]/(f).$$
$$= \mathbb{C}(X), \text{ where } X \subset \mathbb{P}^{n+1} \text{ is the hypersurface with}$$
$$\text{affine equation } f = 0.$$

When $K \cong \mathbb{C}(X)$, and X is a projective variety, we say that X is a *model* of K. Moreover, if X_1, X_2 are 2 models of K, the composition

$$\mathbb{C}(X_1) \xleftarrow{\quad \approx \quad} K \xrightarrow{\quad \approx \quad} \mathbb{C}(X_2)$$

defines a birational correspondence between X_1 and X_2. Thus function fields and birational equivalence classes of varieties are one and the same.

Two of the most fundamental questions in birational geometry are:

i) given K, does it have a *smooth* model X?

ii) given X_1, X_2, 2 smooth models of K, how far is the induced birational map $Z: X_1 \to X_2$ from being biregular? e.g., can it be factored into elementary non-biregular building blocks?

Hironaka [1], building on classic work of Zariski [2], proved that the answer to (i) is always yes. We shall prove this when tr. deg $K = 1$. As for (ii), this is trivial when dim $X = 1$ because of:

(7.1) Proposition. i) *A rational map from a smooth curve X to any variety Y is always regular,*

ii) *A birational map between smooth curves X_1, X_2 is always biregular.*

Proof. (i) comes from Zariski's Main Theorem (3.23) and (ii) from (i).

We will see in Chapter 8 that in dimension 2 a birational map need not be biregular but that it does factor in a canonical and very beautiful way. In dimension 3 and higher, the detailed structure of birational maps is still only imperfectly known.

Once we know that a field K of transcendence degree 1 has a smooth model X, if follows from (7.1) that if Y is any other model, then the birational map

$$Z: X \longrightarrow Y$$

is regular. If $S_Y = \text{Sing } Y$, then S_Y and $S_X = Z^{-1}[S_Y]$ are finite sets and Z gives a biregular map:

$$\text{res } Z : X - S_X \xrightarrow{\;\approx\;} Y - S_Y.$$

Moreover, if $y \in S_Y$, and

$$Z^{-1}[y] = \{x_1, \ldots, x_t\}$$

then for every sufficiently small classical neighborhood U of y, $Z^{-1}[U]$ breaks up into a disjoint union of neighborhoods V_i of x_i, hence U itself breaks up into t *branches*

$$U = \bigcup_{i=1}^{t} Z[V_i]$$

each branch $Z[V_i]$ being a closed analytic subset of U which is homeomorphic to V_i. If $X_1, \ldots, X_n$ are affine coordinates on Y with $y = (0, \ldots, 0)$, and $w_1, \ldots, w_t$ are local analytic coordinates on X at the points $x_1, \ldots, x_t$, then each sufficiently small U can be explicitly parametrized as follows:

(7.2)
$$U = \bigcup_{i=1}^{t} \left\{ \begin{array}{l} \text{image of an analytic map } F_i \\ w_i \to (f_{i1}(w_i), \ldots, f_{in}(w_i)) = F_i(w_i) \\ w_i \in (\text{neigh. } \Delta_i \text{ of } 0 \text{ in } \mathbb{C}) \end{array} \right\}$$

Say for each branch $Z[V_i]$, we write:

$$f_{i1}(w) = a_{i1} w^{r_i} + \text{higher order terms}$$
$$f_{i2}(w) = a_{i2} w^{r_i} + \text{higher order terms}$$
$$\vdots$$
$$f_{in}(w) = a_{in} w^{r_i} + \text{higher order terms}$$

where not all the leading coefficients $a_{i1}, \ldots, a_{in}$ are zero. Then as the tangent cone $E^*_{y,Y}$ is the limit of secants $\overline{y,z}$, $z \to y$, we see

(7.3)
$$E^*_{y,Y} = \bigcup_{i=1}^{t} [\text{line } (a_{i1}t, \ldots, a_{in}t), t \in \mathbb{C}].$$

Moreover, if $\lambda_1, \ldots, \lambda_n$ are such that $\sum_{j=1}^{n} \lambda_j a_{ij} \neq 0$ for all i, then the hyperplane H_0 defined by $\sum \lambda_i X_i = 0$ is not tangent to Y at y. Thus if H_t is defined by $\sum \lambda_i X_i = t$, it follows that for all $t \neq 0$, t sufficiently small:

$$\text{mult}_y Y = \#(U \cap H_t)$$
$$= \sum_{i=1}^{t} \#(Z[V_i] \cap H_t)$$
$$= \sum_{i=1}^{t} \#\{a \in \Delta_i \,|\, \sum_j \lambda_j f_{ij}(a) = t\}.$$

But

$$\sum \lambda_j f_{ij}(w) = (\sum \lambda_j a_{ij})w^{r_i} + \text{higher order terms}$$

so for t small, $t \neq 0, \sum \lambda_j f_{ij}(w) = t$ has r_i zeroes. Thus

(7.4) $$\text{mult}_y \, Y = \sum_{i=1}^t r_i.$$

This is the classical description of a singularity on an algebraic curve.

Returning to the first problem, we next combine the methods of Chapters 5 and 6 to prove:

(7.5) **Theorem.** *For every K of transcendence degree 1, a smooth model exists.*

The following proof is due to Albanese. We start with any model $X \subset \mathbb{P}^k$. Then its Hilbert polynomial is:

$$P_X(t) = \deg X \cdot t + 1 - p_a(X).$$

Let $d = \deg X$ and $\gamma = p_a(X)$ and choose n so that:
a) $\dim(\mathbb{C}[X_0,\ldots,X_k]/I(X))_n = P_X(n)$

b) $\dfrac{nd}{2} > \gamma.$

Now we re-embed X in a higher-dimensional projective space via the composition:

$$X \subset \mathbb{P}^k \xrightarrow{\quad Z_n \quad} \mathbb{P}^N$$

where Z_n is the regular correspondence defined by the complete linear system of hypersurfaces of degree n on $\mathbb{P}^k$, i.e.,

$$Z_n(x_0,\ldots,x_k) = (\ldots.,x^\alpha,\ldots.)_{\text{all } \alpha \text{ with } |\alpha|=n} \,.$$

Let $X_{(0)}$ be the image of X: another model of K. Let $L \subset \mathbb{P}^N$ be the smallest linear space that contains $X_{(0)}$. If we denote the coordinates on $\mathbb{P}^N$ by Y_α, where $\alpha = (\alpha_0,\ldots,\alpha_k)$ is a multi-index such that $|\alpha| = n$ and Z_n is defined by $Y_\alpha = X^\alpha$, then

$$\left(\begin{array}{l}\text{the hyperplane } \sum c_\alpha Y_\alpha \\ \text{contains } X_{(0)}\end{array}\right) \Longleftrightarrow \left(\begin{array}{l}\text{the hypersurface } \sum c_\alpha X^\alpha \\ \text{contains } X\end{array}\right)$$

$$\Longleftrightarrow \sum c_\alpha X^\alpha \in I(X)_n.$$

Therefore

$$\dim L + 1 = \dim\left(\begin{array}{l}\text{vector sp. of all forms } \sum c_\alpha Y_\alpha \\ \text{mod those that vanish on } X_{(0)}\end{array}\right)$$

$$= \dim \mathbb{C}[X_0,\ldots,X_k]_n/I(X)_n$$
$$= P_X(n)$$
$$= nd + 1 - \gamma.$$

Moreover,

$$\deg X_{(0)} = \#\,[X_{(0)} \cap V(\textstyle\sum c_\alpha Y_\alpha)], \text{ if } c_\alpha \text{ generic}$$
$$= \#\,[X \cap V(\textstyle\sum c_\alpha X^\alpha)]$$
$$= \deg X \cdot \deg V(\textstyle\sum c_\alpha X^\alpha) \text{ by Bezout's Th.}$$
$$= nd.$$

In other words, rewriting L as $\mathbb{P}^{nd-\gamma}$, we have a model $X_{(0)} \subset \mathbb{P}^{nd-\gamma}$ of degree nd. If at this point we apply to $X_{(0)}$ the inequality: (5.13) *For all varieties* $V^r \subset \mathbb{P}^n$ *such that* $V \not\subset$ *hyperplane,* $\deg V \geq \operatorname{codim} V + 1$, we get something out already:

(7.6) Corollary. $\gamma \geq 0$, *i.e., if* $X \subset \mathbb{P}^k$ *is any curve, then* $p_a(X) \geq 0$.

However we want to apply (5.13) not only to $X_{(0)}$ but to its projections. Argue as follows:

1) either $X_{(0)}$ is smooth and we are finished; or else $X_{(0)}$ has a singular point x in which case look at the projection $p_x(X_{(0)}) = X_{(1)} \subset \mathbb{P}^{nd-\gamma-1}$.

2) Note that $p_x : X_{(0)} \to X_{(1)}$ is birational. [If not:

$$\deg X_{(1)} \leqq \frac{\deg X_{(0)} - \operatorname{mult}_x X_{(0)}}{2}$$

$$\leqq \frac{nd-2}{2}$$

$$< nd - \gamma - 1$$

$$= \operatorname{codim} X_{(1)} + 1$$

contradicting (5.13)].

3) either $X_{(1)}$ is a smooth model and we are finished; or else $X_{(1)}$ has a singular point x in which case look at the projection $p_x(X_{(1)}) = X_{(2)} \subset \mathbb{P}^{nd-\gamma-2}$. etc.

I claim that eventually some $X_{(k)}$ must be a smooth model. To see this, we also prove by induction that $X_{(k)}$ is birational to $X_{(0)}$. Then since every projection was centered at a singular point, where $X_{(i)}$ had multiplicity ≥ 2:

$$\deg X_{(k)} = \deg X_{(k-1)} - \operatorname{mult}\,(\text{center of proj.})$$

$$\leqq \deg X_{(k-1)} - 2$$

$$\cdots\cdots$$

$$\leqq \deg X_{(0)} - 2k$$

$$= nd - 2k.$$

If $p_x : X_{(k)} \to X_{(k+1)}$ was not birational, looking at $\deg X_{(k+1)}$, we find a contradiction to (5.13) just as in step 2. But it is getting harder and harder for $X_{(k)}$ to satisfy (5.13) because

$$\deg X_{(k)} \leqq nd - 2k \qquad : \text{decreases by 2 in each proj.}$$

$$\operatorname{codim} X_{(k)} + 1 = nd - \gamma - k : \text{decreases by 1 in each proj.}$$

Therefore (5.13) tells us that $k \leq \gamma$: i.e., after at most γ projections we must reach a smooth model.   QED

Even if you start with a smooth $X \subset \mathbb{P}^k$, the above proof gives interesting information:

(7.7) **Corollary.** *If X is a curve and $p_a(X) = 0$, then X is smooth and there is a biregular correspondence between X and $\mathbb{P}^1$.*

Proof. Note that X is smooth if and only if $X_{(0)}$ is smooth. $X_{(0)}$ is a curve of degree d in $\mathbb{P}^d$. If $X_{(0)}$ is singular, then the degree and codimension of $X_{(1)}$ would not satisfy (5.13). If $X_{(0)}$ is smooth, projecting successively from any points of $X_{(0)}$, we continue to get curves of degree k in $\mathbb{P}^k$ birational to X, for successively smaller k. Eventually we get a birational map from $X_{(0)}$ to $\mathbb{P}^1$. Since X and $\mathbb{P}^1$ are smooth, the map must be biregular.   QED

The same sort of argument with curves of degree $d + 1$ in $\mathbb{P}^d$ gives:

(7.8) **Corollary.** *If X is a curve with $p_a(X) = 1$, then either X is singular and is birational to $\mathbb{P}^1$; or X is smooth and admits a biregular correspondence with a non-singular plane cubic.*

§7B. Arithmetic Genus = Topological Genus; Existence of Good Projections to $\mathbb{P}^1, \mathbb{P}^2, \mathbb{P}^3$

It begins to become clear from the above proof how important the arithmetic genus is. For smooth curves, the arithmetic genus has a very simple topological meaning, given by simplest case of the illustrious Hirzebruch-Riemann-Roch theorem. Recall that topologically speaking, all smooth curves are compact oriented 2-manifolds and all these are simply spheres with some number of handles added:

3 handles

The number of handles we call the *topological genus* of the surface. If the surface has g of them, its Betti numbers are

$$B_0 = 1$$

$$B_1 = 2g$$

$$B_2 = 1$$

so its Euler-Poincaré characteristic $e(X)$ is $2 - 2p$. (For the classification of surfaces and numbers, see W. Massey [1].) The general $H - R - R$ theorem for a smooth projective variety X is a formula for $p_a(X)$ in terms of the so-called Chern numbers of X. When $\dim X = 1$, $e(X)$ is the only Chern number that X has and, *mirabile dictu*, the formula reduces to:

(7.9) Hirzebruch-Riemann-Roch theorem in dimension one.   *For all smooth curves X:*

$$\text{arithmetic genus} = \text{topological genus}$$

The most satisfying proof of (7.9) would come from finding a really direct connection between the 2 genera. Such a connection is given by the theory of harmonic forms. However, all presently known proofs of the higher-dimensional $H - R - R$ theorem and most of the proofs of this one-dimensional case depend on reducing the theorem to an explicit case and then making a direct calculation of both sides of the equation. Our proof, which is the classical projective-geometric approach, is of this type and the 2 steps are as follows:

Step I. We investigate *general projections* of a smooth curve $X \subset \mathbb{P}^r$ to $\mathbb{P}^k$. In fact, if L^{-k-1} is a linear space disjoint from X, we get

$$p_k = \text{res } p_L : X \longrightarrow \mathbb{P}^k$$

and:

a) if $k \geq 3$, then for almost all L, $p_k(X)$ is birational to X and is smooth,

b) if $k = 2$, then for almost all L, $p_2(X)$ is birational to X and smooth except for a finite number of ordinary double points or "nodes". These are points $w \in p_2(X)$ of multiplicity 2 whose tangent cone consists in 2 distinct lines; equivalently, the affine equation of $p_2(X)$ in coordinates where $w = (0,0)$ is:

$$(aX + bY)(cX + dY) + \text{(higher order terms)}, \quad ad - bc \neq 0;$$

hence $p_2(X)$ is made up of 2 smooth branches at w crossing transversely (cf. (7.4)),

c) if $k = 1$, then for almost all L of dimension $r - 2$:

$$p_1 : X \longrightarrow \mathbb{P}^1$$

is a covering of degree d smooth except for a finite number of ordinary branch points. These are points $P \in X$ where $\text{mult}_P(p_1) = 2$; or equivalently if x is a local coordinate on X near x, t a local coordinate on $\mathbb{P}^1$ near $p_1(x)$, then

$$t = ax^2 + \text{(higher order terms)}, a \neq 0.$$

Generalizations of Step I to higher-dimensional cases have been much studied. For instance, if X is a smooth surface and we project it suitably into $\mathbb{P}^3$, then it can be shown that its singularities, in suitable local analytic coordinates x, y, z are of the 3 types:

$$x \cdot y = 0 \text{ ("ordinary double curve")}$$

$$x \cdot y \cdot z = 0 \text{ ("ordinary triple point")}$$

$$x^2 - y \cdot z^2 = 0 \text{("pinch point")}.$$

The third type is a surface with the double curve $y : x = z = 0$, and at all points $y \neq 0$, 2 non-singular "branches" $x = +\sqrt{y \cdot z}$, $x = -\sqrt{y \cdot z}$ through y; but where the 2 branches are interchanged by going around the loop $y = e^{i\theta}$ in γ. The reader may enjoy visualizing this situation, also known as "Cartan's umbrella".

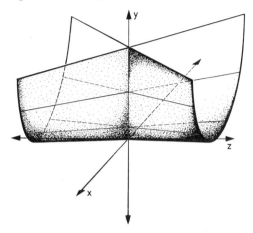

Step II. If $d = \deg(X)$, $e(X) =$ Euler-characteristic of X, $p_2(X)$ has v nodes and $p_1 : X \to \mathbb{P}^1$ has β branch points, then we will prove:

$$a) \qquad p_a(X) = \frac{(d-1)(d-2)}{2} - v,$$

(7.10)
$$b) \quad d(d-1) = \beta + 2v$$

$$c) \qquad e(X) = 2d - \beta.$$

A little algebra shows that these imply: topological genus $(X) = p_a(X)$.

Proof of Step I. The 1st part of this step will consist in showing that given $X \subset \mathbb{P}^k$, if $k > 3$, then for almost all $x \in \mathbb{P}^k$, $p_x(X)$ is still smooth and birational to X. In fact, it suffices to find $x \in \mathbb{P}^k - X$ such that:

(7.11) $$\qquad for \; all \; u \in X, \overline{ux} \cap X = \{u\} \; and \; \overline{ux} \neq E^*_{u,X}.$$

If x satisfies (7.11), then

$$\mathrm{res} \; p_x : X \longrightarrow p_x(X) = X'$$

is bijective, hence birational, and by (5.14), all points of X' have multiplicity 1, hence X' is smooth. On the other hand, for x to satisfy (7.11), it suffices that

$$x \notin \underbrace{\left[\bigcup_{\substack{u \neq v \\ u,v \in X}} \overline{uv} \right] \cup \left[\bigcup_{u \in X} E^*_{u,X} \right]}_{S}$$

But I claim that S is an algebraic variety of dimension ≤ 3, so such an x exists. To see this, look at

$$C \subset X \times X \times \mathbb{P}^k$$

$$C = \{(x,y,z) \,|\, \text{if } x \neq y, \text{ then } z \in \overline{xy}; \text{ if } x = y, \text{ then } z \in E^*_{x,X}\}.$$

$C \cap (X \times X - \Delta) \times \mathbb{P}^k$ is obviously a closed algebraic set, and the whole of C is the closure of this open piece in the classical topology (since if $x_n, y_n \to x$ as $n \to \infty$, then the secant $\overline{x_n y_n}$ approaches the tangent line $E^*_{x,X}$). Therefore C is a closed algebraic set. But the fibres of $p_{12} : C \to X \times X$ are all lines which are irreducible 1-dimensional. Therefore C is irreducible 3-dimensional. Since $p_3(C) = S$, S is irreducible and (≤ 3)-dimensional.

Now say $X \subset \mathbb{P}^3$ is a smooth curve. To prove part (b) of Step I, we seek $x \in \mathbb{P}^3$ such that:

i) $x \notin \bigcup_{u \in X} E^*_{u,X}$

ii) x is in only a *finite* set of secants $\overline{u_i v_i}$ of X, $1 \leq i \leq v$

(7.12)   iii) for all i, $\overline{u_i v_i}$ is not a tri-secant of X

iv) for all i, the 2 planes $\overline{x, E^*_{u_i,X}}$ and $\overline{x, E^*_{v_i,X}}$ are distinct.

Given such an x, let $X' = p_x(X) \subset \mathbb{P}^2$. Then (ii) means that X' is birational to X and by (i), its only singular points are $w_i = p_x(u_i) = p_x(v_i)$.

Now a small neighborhood of w_i in X' is the union of images of neighborhoods of u_i and v_i in X; each of these pieces is an analytic set through w_i which by (i) and the analytic implicit function theorem is a smooth 1-dimensional analytic manifold through w_i. Therefore, almost all lines $l \subset \mathbb{P}^2$ passing near w_i meet each of these branches in 1 point near w_i, hence meet X' in 2 points near w_i; hence the multiplicity of w_i on X' is 2. By (iv), these 2 branches have distinct tangent lines, namely $p_x(E^*_{u_i,X})$ and $p_x(E^*_{v_i,X})$. Therefore the tangent cone to X' at w_i consists of these 2 distinct lines, and w_i is an ordinary double point of X'.

We must show that such an x exists. Introduce 2 subsets of $X \times X - \Delta$:

$$T = \text{set of ``trisecants'' in } X \times X - \Delta$$

$$= \left\{ (x,y) \middle| \begin{array}{l} \text{either } \overline{xy} \text{ meets } X \text{ in a 3rd point} \\ x \neq y \text{ or } \overline{xy} \text{ is tangent to } X \text{ at } x \text{ or } y \end{array} \right\}$$

$$B = \text{set of secants in bitangent planes}$$

$$= \left\{ (x,y) \middle| \begin{array}{l} \text{the 2 lines } E^*_{x,X}, E^*_{y,X} \text{ lie in a plane} \\ x \neq y \end{array} \right\}$$

$$T \qquad\qquad\qquad\qquad\qquad B$$

It is easy to see that T and B are closed algebraic subsets of $X \times X - \Delta$. We prove next that:

(*) $$T \cup B \subsetneqq (X \times X - \Delta).$$

Construction of a "good" secant. Start with any $x \in X$ and let $l = E^*_{x,X}$. Look at the projection

$$\text{res } p_l : X \longrightarrow \mathbb{P}^1.$$

Let $\alpha \in \mathbb{P}^1$ be a point over which res p_l is smooth and let $p_l^{-1}(\alpha) \cup l = L$, $L \cap X = \{x, y_1, \ldots, y_k\}$. But if $y \notin l$, res p_l is smooth at y means that $p_l(E^*_{y,X}) = \mathbb{P}^1$. Therefore $E^*_{y_i,X} \not\subset L$, i.e., $(x, y_i) \notin B$:

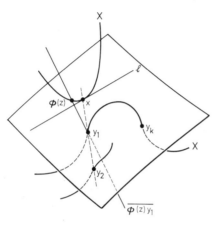

Let

$$\phi : \{z \,||z| < \varepsilon\} \longrightarrow X$$
$$\phi(0) = x$$

be an analytic coordinate on X near x. Consider the secants $\overline{y_1 \phi(z)}$. Since $(x, y_1) \notin B$, $(\phi(z), y_1) \notin B$ if $|z|$ is small. Now perhaps $(x, y_1) \in T$, i.e., x, y_1 and some y_i's, $2 \le i \le k$, are collinear in L. But if $(\phi(z), y_1) \in T$ for all small z, there would be an i, $2 \le i \le k$ and sequences

$$\left. \begin{array}{l} z_n \longrightarrow 0 \\ y_i^{(n)} \longrightarrow y_i, y_i^{(n)} \in X \end{array} \right\} \text{ as } n \to \infty$$

such that

$$\phi(z_n), y_1, y_i^{(n)} \text{ collinear.}$$

Then the secant $\overline{y_i y_i^{(n)}}$ to X would lie in the plane $\overline{y_1 x \phi(z_n)}$. As $n \to \infty$, this plane approaches the join of y_1 and $E^*_{x,X}$, i.e., L. Therefore the tangent line $E^*_{x,X}$ would lie in L — a contradiction. Thus $(\phi(z), y_1) \in X \times X - \Delta - B - T$ for some small z.

Finally, going back to the notation of the 1st part of the proof, let

$$C^* = C \cap [(\Delta \cup B \cup T) \times \mathbb{P}^3]$$
$$S^* = p_3(C^*).$$

Then C^* is a closed algebraic set all of whose components have dimension ≤ 2, hence the same holds for S^*. But by definition S^* is the union of the tangent lines to X, the trisecants and the secants which lie in bitangent planes. If we choose $x \notin S^*$, it has all the required properties.

Now for part (c) of Step I say $X \subset \mathbb{P}^2$ is a plane curve with only ordinary double points. If $x \in X$ is a smooth point, then we say x is a *point of inflexion* if

$$i(x; X \cap E^*_{x,x}) \geqq 3.$$

The points of inflexion can be found by various methods: viz.

(7.13) **Proposition.** *Let X be a plane curve.*
1) *if $f(X_0, X_1, X_2) = 0$ is its homogeneous equation, then:*

$$\left\{\begin{array}{c} \text{points of} \\ \text{inflexion} \\ \text{on } x \end{array}\right\} = (X - \text{Sing } X) \cap \left\{\begin{array}{c} \text{zeroes of the Hessian} \\ H = \det\left(\dfrac{\partial^2 f}{\partial X_i \partial X_j}\right) \end{array}\right\}$$

2) *if X is described analytically in affine coordinates near P by $X_2 = f(X_1)$, then*

$$\left\{\begin{array}{c} \text{points of} \\ \text{inflexion} \\ \text{near } P \end{array}\right\} = \left\{\begin{array}{c} \text{points } (a, f(a)) \\ \text{where } f''(a) = 0 \end{array}\right\}.$$

Proof. This is an exercise in 2nd year calculus. (1), for instance, can be proved like this. First one checks that the Hessian curve $H = 0$ is independent of which coordinates are used to calculate it in. So choose coordinates such that $x = (1,0,0)$, $E^*_{x,x} = V(X_1)$. Then the equation f of X looks like:

$$f = \alpha X_1 X_0^{d-1} + \beta X_2^2 X_0^{d-2} + \gamma X_1 X_2 X_0^{d-2} + \delta X_1^2 X_0^{d-2}$$

$$+ \text{(terms with } X_1, X_2 \text{ to powers} > 2).$$

Calculating, one finds:

$$H(1,0,0) = -2(d-1)^2 \alpha^2 \beta.$$

Then $\alpha \neq 0$ if x is a smooth point of X, hence $H(x) = 0$ if and only if $\beta = 0$. But

$$X \cap E^*_{x,x} = \text{zeroes of } f(X_0, 0, X_2)$$
$$= \text{zeroes of } \beta X_2^2 X_0^{d-2} + \text{(terms with } X_2 \text{ to higher power)}$$

hence $\beta = 0$ if and only if the order to contact of $X, E^*_{x,x}$ is bigger than 2. We leave (2) to the reader.   QED

(7.14) **Corollary.** *If $\deg X > 1$, X has only finitely many points of inflexion.*

Proof. By (2), since X is not a line, $f'' \not\equiv 0$ and hence not *every* point on X is a point of inflexion. By (1), it is an algebraic set, hence is finite.

Now returning to our curve X with ordinary double points, choose $x \in \mathbb{P}^2$ such that

$$x \notin \bigcup_{\substack{y \in Xa \\ \text{point of} \\ \text{inflexion or} \\ \text{a singular point}}} E^*_{y,X} \cup \dot{X}$$

Consider $p_x : X \to \mathbb{P}^1$. Then I claim that at every smooth point $y \in X, p_x$ has at most an *ordinary* branch point; and at every ordinary double point, p_x restricted to either branch of X is a smooth map. Combined with our previous projections, this will prove part (c) of Step I. In fact, for all smooth points $y \in X$, if $x \notin E_{y,X}^*$, then p_x is smooth at y; and at the double points y, since x is not on the tangent line to either smooth branch, p_x is smooth on each branch. Now if y is smooth, and $x \in E_{y,X}^*$, then y is not a point of inflexion. Therefore

$$2 = i(y; X \cap E_{y,X}^*)$$

$$= i(y; X \cap p_x^{-1}(p_x(y)))$$

$$= \text{mult}_y(\text{res}_X p_x).$$

In other words, for all $z \in \mathbb{P}^1$ near $p_x(y)$, $p_x^{-1}(z) \cap X$ contains 2 points near y. This means y is an ordinary branch point.

Proof of Step IIa. Consider the birational projection

$$X \xrightarrow{\quad p_2 \quad} X' = V(F) \subset \mathbb{P}^2$$

$$u_i \mapsto$$
$$\searrow w_i \qquad 1 \le i \le v.$$
$$v_i \mapsto$$

and where X' is smooth at all points except the w_i. Then $X - \{u_1, v_1, \ldots, u_v, v_v\}$ and $X' - \{w_1, \ldots, w_v\}$ correspond biregularly to each other under p_2 and in particular if $x \in X - \{u_1, \ldots, v_v\}$ and $x' = p_2(x)$, then identifying $\mathbb{C}(X)$ with $\mathbb{C}(X')$

$$\mathcal{O}_{x,X} = \mathcal{O}_{x',X'}.$$

The main point in comparing the Hilbert polynomials of X and X' is to compare the local rings at the singular points:

(7.15) Lemma.　　　$\mathcal{O}_{w_i, X'} = \{\alpha \in \mathcal{O}_{u_i, X} \cap \mathcal{O}_{v_i, X} \mid \alpha(u_i) = \alpha(v_i)\}.$

Proof. We use the approximation technique of Chapter 1. Let Y, Z be affine coordinates in $\mathbb{P}^2$ such that $w_i = (0,0)$ and such that the affine equation F of X' is of the form:

$$F = Y \cdot Z + \text{higher order terms.}$$

Also, recall that for every n every function $f \in \mathcal{O}_{w_i, \mathbb{P}^2}$ can be expanded

$$f = \sum_{i+j<n} c_{ij} Y^i Z^j + (\text{remainder in } \mathfrak{M}_{w_i}^n{}^2).$$

Looking at this modulo F, it follows that every function $f \in \mathcal{O}_{w_i, X'}$ can be expanded

(*)　　　　$f = a + \sum_{i=1}^{n-1} b_i Y^i + \sum_{i=1}^{n-1} c_i Z^i + (\text{remainder in } \mathfrak{M}_{w_i, X'}^n).$

As an analytic set near w_i, X' is the union of 2 smooth branches with tangent lines $Y = 0$ and $Z = 0$, which are isomorphic under p_2 with small neighborhoods of u_i and v_i on X. On the branch with tangent line $Y = 0$, Z vanishes to 1st order and Y to higher order, so if this branch corresponds to a neighborhood of u_i on X, the function $Y \circ p_2$ vanishes to 1st order at u_i, and $Z \circ p_2$ to higher order. At v_i, the opposite happens. From this it follows immediately that any function $f \in \mathfrak{M}_{u_i,X} \cap \mathfrak{M}_{v_i,X}$ can be expanded:

$$(**) \qquad f = \sum_{i=1}^{n-1} b_i Y^i + \sum_{i=1}^{n-1} c_i Z^i + \left(\begin{array}{c} \text{remainder vanishing to} \\ \text{order } n \text{ at } u_i \text{ and } v_i \end{array} \right).$$

Now start with $f \in \mathcal{O}_{u_i,X} \cap \mathcal{O}_{v_i,X}$ such that $f(u_i) = f(v_i) = 0$. Write $f = g/h$ with $g, h \in \mathcal{O}_{w_i,X'}$. For all n expand f as in (**):

$$f = f_n(Y,Z) + R_n$$

$$f_n \text{ polynomial of degree } n - 1$$

$$R_n \text{ vanishing to order } n \text{ at } u_i \text{ and } v_i.$$

Then

$$g = h \cdot f_n + h \cdot R_n.$$

Since g and $h f_n \in \mathcal{O}_{w_i,X'}$, $h R_n \in \mathcal{O}_{w_i,X'}$ also. If we expand $h R_n$ as in (*), it follows that all the coefficients a, b_n, c_n are 0 or else $h R_n$ would not vanish to order n at u_i and v_i. Therefore $h R_n \in \mathfrak{M}^n_{w_i,X'}$. Thus for all n

$$g \in h \cdot \mathcal{O}_{w_i,X'} + \mathfrak{M}^n_{w_i,X'}.$$

Therefore by Krull's theorem, $g \in h \cdot \mathcal{O}_{w_i,X'}$, i.e., $f \in \mathcal{O}_{w_i,X'}$.   QED

We now proceed to compare the 2 homogeneous coordinate rings:

$$R = \mathbb{C}[X_0, \ldots, X_r]/I(X)$$

$$\cup$$

$$R' = \mathbb{C}[X_0, X_1, X_2]/(F).$$

We may assume after a suitable change of coordinates that $X_i(w_j) \neq 0$ $i = 0, 1, 2$, $j = 1, \ldots, v$. For all k, consider the sequence:

$$0 \longrightarrow R'_k \xrightarrow{\ i\ } R_k \xrightarrow{\ \alpha\ } \sum_{i=1}^{v} \mathbb{C} \longrightarrow 0$$

$$(7.16)_k$$

$$\alpha(f) = \left(\ldots \ldots, \frac{f}{X_0^k}(u_i) - \frac{f}{X_0^k}(v_i), \ldots \ldots \right)$$

$$i = \text{given inclusion}.$$

It is clear that $\alpha \circ i = 0$. *I claim that* $(7.16)_k$ *is exact if* $k \gg 0$. There are 2 things to check here: the first is quite simple: namely that α is surjective. In fact if $k \gg 0$, there are hypersurfaces $H_i = V(G_i)$ such that

$$u_j, v_j \in H_i \quad \text{all } j \neq i$$

$$u_i \notin H_i, \ v_i \in H_i.$$

Then $\alpha(G_i) = $ const. (*i*th unit vector), so α is surjective. The second is that if $G \in R_k$ and $\alpha(G) = 0$, then $G \in R'_k$. To prove this we can apply (6.11). Since p_2 is birational, R and R' have the same quotient field and it therefore suffices to prove that

(7.17) **Lemma.** $\forall\, k$ and $\forall\, G \in R_k$ such that $\alpha(G) = 0$, $\exists\, n$ such that

$$X_0^n G, X_1^n G, X_2^n G \in R'.$$

By symmetry it will suffice to prove $X_0^n G \in R'$. This is now equivalent to a statement relating the affine rings of $X - X \cap V(X_0)$, $X' - X' \cap V(X_0)$:

$$S = \mathbb{C}\left[\frac{X_1}{X_0}, \ldots, \frac{X_r}{X_0}\right] \bigg/ I(X)$$

$$S' = \mathbb{C}\left[\frac{X_1}{X_0}, \frac{X_2}{X_0}\right] \bigg/ \left(\frac{F}{X_0^d}\right)$$
$$\| \qquad\qquad$$
$$f \qquad\qquad$$

Namely:

(7.18) **Lemma.** If $g \in S$ is such that $g(u_i) = g(v_i)$, all i, then $g \in S'$.

But by (1.11) $S = \bigcap\limits_{\substack{x \in X \\ x \notin V(X_0)}} \mathcal{O}_{x,X}$ and $S' = \bigcap\limits_{\substack{x \in X' \\ x \notin V(X_0)}} \mathcal{O}_{x,X'}$, so this follows from Lemma (7.15) and the equality of all other local rings of X and X'.

Now by (7.16)$_k$, if $k \gg 0$ and $d = \deg X = \deg F$:

$$P_X(k) = \dim R_k$$
$$= \dim[\mathbb{C}[X_0, X_1, X_2]/(F)]_k + v$$
$$= \dim \mathbb{C}[X_0, X_1, X_2]_k - \dim F \cdot \mathbb{C}[X_0, X_1, X_2]_{k-d} + v$$
$$= \frac{(k+1)(k+2)}{2} - \frac{(k-d+1)(k-d+2)}{2} + v$$
$$= kd + 1 - \frac{(d-1)(d-2)}{2} + v,$$

hence

$$p_a(X) = \frac{(d-1)(d-2)}{2} - v.$$

Proof of Step IIb. This is an application of Bezout's Theorem. In fact, let X_0, X_1, X_2 be coordinates in the $\mathbb{P}^2$ into which X is projected, let X_0, X_1 be coordinates in $\mathbb{P}^1$ and let

$$p_x : \mathbb{P}^2 - \{x\} \longrightarrow \mathbb{P}^1$$

be the projection. Let $F(X_0, X_1, X_2) = 0$ be the equation of $p_2(X) \subset \mathbb{P}^2$. Consider the auxiliary curve Y defined by $\partial F/\partial X_0 = 0$. Y has degree $d - 1$, so by Bezout's theorem

$$\deg(Y \cdot p_2 X) = d(d-1).$$

I claim:

    a) $Y \cap p_2 X = $ (double points of $p_2 X$ and branch points

$$\text{of } \operatorname{res} p_x : p_2 X \longrightarrow \mathbb{P}^1)$$

    b) at each double point $y \in p_2 X$,

$$i(y; Y \cap p_2 X) = 2,$$

    c) at each branch point $y \in p_2 X$ of $p_2 X \to \mathbb{P}^1$,

$$i(y; Y \cap p_2 X) = 1,$$

which gives us the formula

$$d(d-1) = 2v + \beta.$$

To prove these, since $p_2 X$ is covered by affine pieces $X_1 \neq 0$ and $X_2 \neq 0$, look in the first piece and let $u = X_0/X_1, v = X_2/X_1$ be affine coordinates. Then p_x becomes projection of the (u,v)-plane to the u-line, $p_2 X$ has an affine equation $f(u,v) = F(u,1,v)$ and Y has an affine equation $\dfrac{\partial F}{\partial X_0}(u,1,v)$ which is just $\partial f/\partial u$.

Then for all $y \in p_2 X$,

$$\frac{\partial f}{\partial u}(y) \neq 0 \Longleftrightarrow \begin{array}{l} y \text{ is a smooth point of } X \text{ and} \\ E^*_{y,p_2 X} \text{ projects onto the } u\text{-line} \end{array}$$

hence (a) follows. If y is a double point and the coordinates (u,v) are chosen with $y = (0,0)$, then because neither branch of $p_2 X$ at y is parallel to the v-axis:

$$f = (au + bv) \cdot (cu + dv) + \deg \geq 3 \text{ terms}, \quad ad - bc \neq 0, a \neq 0, c \neq 0$$

hence

$$\frac{\partial f}{\partial u} = 2ac \cdot u + (ad + bc)v + (\deg \geq 2 \text{ terms}).$$

Since

$$\frac{ad + bc}{2ac} \neq \frac{b}{a} \text{ or } \frac{d}{c}$$

Y has a smooth point at y with tangent line unequal to either tangent line to $p_2 X$ at y. So we get (b). Finally, if y is a branch point of $\operatorname{res} p_x : p_2 X \to \mathbb{P}^1$, and the coordinates (u,v) are chosen with $y = (0,0)$ again, then because y is smooth and the branching is ordinary:

$$f = av + bu^2 + cuv + dv^2 + (\deg \geq 3 \text{ terms}), \quad a \neq 0, b \neq 0$$

hence

$$\frac{\partial f}{\partial u} = 2bu + (\deg \geq 2 \text{ terms}).$$

Thus y is a smooth point of Y and Y and $p_2 X$ meet transversely at y. This proves (c).

Proof of Step IIc. Here we merely examine the covering $p_1 : X \to \mathbb{P}^1$. Let us simultaneously triangulate X and $\mathbb{P}^1$ and calculate $e(X)$ as the number of vertices minus the number of edges plus the number of faces! Let $x_1, \ldots, x_\beta \in X$ be the branch points of X and let $t_i = p_1(x_i)$. Then start with any triangulation of $\mathbb{P}^1$ which includes the t_i among its vertices. Then triangulate X as follows: take all points of X over vertices of $\mathbb{P}^1$ to be vertices. For all edges

$$f : \Delta^1 \longrightarrow \mathbb{P}^1$$

in the triangulation, the covering X is unramified over $f(\text{Int } \Delta^1)$, and since $\text{Int } \Delta^1$ is simply connected, f lifts to d distinct maps

$$f_0^{(i)} : \text{Int } \Delta^1 \longrightarrow X$$

with disjoint images. Since $p_1 : X \to \mathbb{P}^1$ is proper, we can then extend $f_0^{(i)}$ to maps

$$f^{(i)} : \Delta^1 \longrightarrow X$$

lifting f. Let these be the edges of the triangulation of X. Do the same with faces. In this way, we wind up with a triangulation of X. Now if the triangulation of $\mathbb{P}^1$ had s_0 vertices, s_1 edges and s_2 faces, that of X has ds_1 edges and ds_2 faces since p_1 was unramified over these. But among the ds_0 potential points of X over the t_i, β are branch points where 2 sheets have come together. So there are $ds_0 - \beta$ vertices. Thus

$$e(X) = (ds_0 - \beta) - (ds_1) + (ds_2)$$
$$= d(s_0 - s_1 + s_2) - \beta$$
$$= d(e(\mathbb{P}^1)) - \beta$$
$$= 2d - \beta.$$

This proves Step IIc and Theorem (7.9).   QED

Theorem (7.9) has various Corollaries. For instance, combined with (7.7), it gives:

(7.17) Corollary. *If X is a smooth curve, the following are equivalent:*

   i) $p_a(X) = 0$
   ii) *X is homeomorphic in the classical topology to a sphere S^2,*
   iii) *X is in biregular correspondence with $\mathbb{P}^1$.*

Using the concept of the canonical class (§6B), and, in particular, (6.19), we get a more or less equivalent reformulation of (7.9). Recall from (3.27) that if on a smooth curve X, we define

$$\deg\left(\sum n_i x_i\right) = \sum n_i$$

for all $\sum n_i x_i \in \text{Div}(X)$, then degree depends only on the linear equivalence class of the divisor. In particular, all canonical divisors (ω) have the same degree, that we denote $\deg(K)$. Then:

(7.18) Corollary. *If X is a smooth curve,*

$$\deg K = 2p_a(X) - 2.$$

Proof. If $p_1 : X \to \mathbb{P}^1$ is a projection as in the Proof of (7.9), then by (6.19):

$$\deg K = \deg(\text{branch locus}) - 2 \deg(\text{hyperplane section})$$
$$= \beta - 2d.$$

But by Step IIa and IIb, one checks immediately:

$$\beta - 2d = 2p_a(X) - 2. \quad \text{QED}$$

In fact to go back from (7.18) to the Theorem (7.9), one would need only to check directly that

(7.19) $$\deg K = -e(X).$$

This is not hard with a little elementary differential topology (as in, for instance, Milnor [3]). For instance, if ω is any rational 1-form on X one can define a vector field D on X outside the zeroes and poles of ω as follows: for all $x \in X$, let $D_x \in T_{x,X}$ be the solution to $\omega(D_x) = 1$. This vector field has singularities at the zeroes and poles of ω, but one can check that at each such point x,

$$(\text{index of } D \text{ at } x) = -\text{ mult. of } x \text{ in } (\omega).$$

Since the sum of the indices of D at all its singularities is the Euler characteristic, we get (7.19).

Another Corollary is the Hurwitz formula:

(7.20) **Corollary.** *If $\pi : X \to Y$ is a regular surjective map of smooth curves of degree d and B is the branch locus divisor on X, then*

$$2 p_a(X) - 2 = \deg B + d(2 p_a(Y) - 2).$$

In particular,

$$p_a(X) \geq p_a(Y).$$

Proof. Combine (7.18) and (6.20).

§7C. Residues of Differentials on Curves, the Classical Riemann-Roch Theorem for Curves and Applications

In this section, we will apply some ideas of complex function theory of 1 variable to derive the "classical Riemann-Roch theorem": this is a simple and explicit formula for dim $\mathcal{L}(D)$ for any divisor D on a smooth curve X, and is the central result in the whole higher theory of curves and functions on curves. To be specific, what we need is the definition and first properties of the concept of *residue*.

Recall that if ω is a meromorphic differential (or 1-form) defined in some neighborhood $U \subset \mathbb{C}$ of the origin, then ω can be expanded:

$$\omega = \left(\frac{a_{-n}}{z^n} + \frac{a_{-n+1}}{z^{n-1}} + \ldots + \frac{a_1}{z} + a_0 + a_1 z + a_2 z^2 + \ldots \right) \cdot dz$$

and if $V \ll U$ is a relatively compact open subset of U with differentiable boundary containing 0 and no other singularities of ω, then

(7.21)
$$\int_{\partial V} \omega = 2\pi i \cdot a_{-1}.$$

One reason this is remarkable is that it shows that the coefficient a_{-1} can be expressed in a coordinate-invariant fashion as a contour integral, hence it does not change if we make any change in coordinates of the form:

$$z' = b_1 z + b_2 z^2 + \dots, b_1 \neq 0.$$

The coefficient a_{-1} is called the residue of ω at 0, or $\mathrm{res}_0 \omega$. It follows that if X is any 1-dimensional complex manifold and ω is any meromorphic differential on X, then for all singularities $x \in X$ of ω, we can define

$$\mathrm{res}_x \omega = \frac{1}{2\pi i} \int_{\partial V} \omega \quad \text{or} = a_{-1}$$

either for any relatively compact open neighborhood V of x, with differentiable boundary, contained in a coordinate neighborhood of x and not containing any other singularities of ω; or via the coefficient of $\dfrac{dz}{z}$ in the expansion of ω in any local coordinate z. But now suppose X is a *compact* l-dimensional complex manifold and that ω is a meromorphic differential on X. Then the fundamental result is that:

(7.22)
$$\sum_{\substack{\text{singularities} \\ x \text{ of } \omega}} \mathrm{res}_x \omega = 0$$

Both formulae (7.21) and (7.22) are simple consequences of Stoke's theorem.
To prove (7.21), note that when V is a small disc centered at the origin, then (7.21) is immediate by an explicit calculation in polar coordinates $z = re^{i\theta}$:

$$\int_{|z|=\varepsilon} z^k dz = i\varepsilon^{k+1} \int_0^{2\pi} e^{(k+1)i\theta} d\theta = \begin{cases} 0, & k \neq -1 \\ 2\pi i, & k = -1. \end{cases}$$

In general, let $V_0 \ll V$ be such a small disc, and let $U = V - \overline{V_0}$. Then because ω is a *closed* differential form

$$\int_{\partial U} \omega = \int_U d\omega = 0.$$

But $\partial U = \partial V - \partial V_0$, hence

$$\int_{\partial V} \omega = \int_{\partial V_0} \omega = 2\pi i \cdot a_{-1}.$$

To prove (7.22), let $x_1, \dots, x_n$ be the singularities of ω (there are only finitely many because they form a discrete subset of a compact space X). Let V_i be nice disjoint small neighborhoods of x_i as above. Let

$$U = X - \bigcup_{i=1}^{n} \bar{V}_i.$$

Then as above we have

$$\int_{\partial U} \omega = \int_{U} d\omega = 0,$$

hence

$$\sum_{i=1}^{n} \operatorname{res}_{x_i} \omega = \sum_{i=1}^{n} \frac{1}{2\pi i} \int_{\partial V_i} \omega = -\frac{1}{2\pi i} \int_{\partial U} \omega = 0.$$

The use we want to make of (7.22) is to give obstructions to the construction of meromorphic functions with prescribed poles. Start now* with a smooth algebraic curve X and points $x_1, \ldots, x_n$. Let $z_i \in \mathcal{O}_{x_i,X}$ be local analytic coordinates at each x_i with $z_i(x_i) = 0$. We ask the classical *Mittag-Loeffler problem*:

(7.23) *given polar parts* $f_i = \sum_{k=-1}^{-d_i} a_{ik} z_i^k$, *is there a rational function f on X, with poles only at* $\{x_1, \ldots, x_n\}$ *such that near* x_i:

$$f = f_i + \sum_{k=0}^{\infty} a_{ik} \cdot z_i^k,$$

i.e., $f - f_i$ analytic?

Note that if f exists, it is unique up to adding a constant function (because, if f, f' were 2 solutions, then $f - f'$ would be a rational function with no poles, hence a constant). When $X = \mathbb{P}^1$, i.e., genus 0, one constructs a solution easily by partial fractions. Namely, let t be an affine coordinate on $\mathbb{P}^1$ such that no x_i is the point $t = \infty$. Let $t(x_i) = a_i$. Then we may choose $z_i = t - a_i$, and the rational function:

$$f = \left[\frac{a_{1,-1}}{t - a_1} + \cdots + \frac{a_{1,-d_1}}{(t - a_1)^{d_1}} \right] + \cdots\cdots + \left[\frac{a_{n,-1}}{t - a_n} + \cdots + \frac{a_{n,-d_n}}{(t - a_n)^{d_n}} \right]$$

solves the problem. However, suppose the curve X has everywhere regular 1-forms ω (= rational 1-forms with no poles). Then we get as follows a linear equation on the coefficients a_{ik} which must be satisfied if f is to exist:

(7.24) *expand near* x_i:

$$\omega = \sum_{k=0}^{\infty} b_{ik} z_i^k \cdot dz_i.$$

---

*Actually, all compact 1-dimensional complex manifolds are curves, but we don't need to use this (see §4B).

Then if f exists:

$$0 = \sum_{i=1}^{n} \operatorname{res}_{x_i}(f\omega)$$

$$= \sum_{i=1}^{n} \operatorname{res}_{x_i}(f_i\omega) + \sum_{i=1}^{n} \operatorname{res}_{x_i} \overbrace{(f-f_i)\cdot\omega}^{\text{no pole}}$$

$$= \sum_{i=1}^{n} \text{coeff. of } z_i^{-1} \text{ in } \left(\sum_{k=-1}^{-d_i} a_{ik}z_i^k \right) \cdot \left(\sum_{k=0}^{\infty} b_{ik}z_i^k \right)$$

$$= \sum_{i=1}^{n} \sum_{k=1}^{d_i} a_{i,-k}\cdot b_{i,k-1}.$$

The classical Riemann-Roch theorem tells us that, conversely if this equation is satisfied for all everywhere regular ω, then f exists. We will prove this shortly. First note that it may be expressed purely numerically by the formula:

$$\dim[\ \mathcal{L}(d_1x_1 + \ldots + d_nx_n)/\text{constant fcns.}]$$

$$= (d_1 + \ldots + d_n) - \begin{bmatrix} \text{number of lin. ind. cond.} \\ \text{given by } \omega \in \Omega_X^1 \end{bmatrix}$$

But a differential $\omega \in \Omega_X$ gives no restriction on the $\{a_{ij}\}$ if and only if $b_{i,k-1} = 0$, $1 \leq k \leq d_i$. This means that ω has a *zero* of order at least d_i at x_i, or that:

$$\omega \in \Omega_X(-d_1x_1 - \ldots - d_nx_n).$$

Writing D for the divisor $d_1x_1 + \ldots + d_nx_n$, we come up with the equivalent formula:

$$\dim \ \mathcal{L}(D) - 1 = \deg D - \dim \Omega_X/\Omega_X(-D)$$

or (in the terminology of (6.15)):

(7.25) $$\dim \ \mathcal{L}(D) - \dim \ \mathcal{L}(K_X - D) = \deg D - p_g(X) + 1.$$

We now state in its full force:

(7.26) **Classical* Riemann-Roch formula for curves.** *Let X be a smooth projective curve. Then* $p_g(X) = p_a(X) = \dfrac{\deg K_X}{2} + 1$ *and we call this number simply the genus g; for all divisors D on X, we have:*

$$\dim \ \mathcal{L}(D) - \dim \ \mathcal{L}(K_X - D) = \deg D - g + 1.$$

Proof. We start with a variant of the above residue argument to give us an inequality. Let D be any divisor. For all sufficiently large m, we can find a divisor $D_m \in |mH|$ such that $D_m \geq D$. (In fact, for all $x \in X$, let H_x be a hyperplane through x and not containing X identically and let $D_x = X \cdot H_x \in |H|$, so that x is in the

---

*We call this "classical" first of all because it it the original Riemann-Roch theorem, and secondly because it is not the dimension 1 case of the Hirzebruch-Riemann-Roch theorem but rather the combination of this case of $H - R - R$ plus the dimension 1 case of "Serre duality".

support of D_x. Then if $D = \sum\limits_{i=1}^{k} u_i - \sum\limits_{i=1}^{l} v_i$, let $D_m = \sum D_{u_i} + D'$, any $D' \in |(m-k)H|$.)

Fix some m and write out:

$$D = d_1 x_1 + \ldots + d_n x_n$$

$$D_m = e_1 x_1 + \ldots + e_n x_n$$

$$e_i \geqq d_i.$$

Let z_i be a local coordinate at x_i and let V be the vector space of formal expansions

$$a_{1,d_1+1} z_1^{-(d_1+1)} + \ldots\ldots + a_{1,e_1} z^{-e_1}$$

$$\ldots\ldots$$

$$a_{n,d_n+1} z_n^{-(d_n+1)} + \ldots\ldots + a_{n,e_n} z_n^{-e_n}.$$

Note that there is a natural map

$$\mathscr{L}(D_m) \longrightarrow V$$

given by mapping a rational function $f \in \mathscr{L}(D_m)$ to the set of leading pieces of its power series expansions at the x_i: more precisely, at x_i, we can write

$$f = \sum_{k=-e_i}^{+\infty} a_{i,-k} \cdot z_i^k$$

and we take the top $e_i - d_i$ terms of this expansion. The kernel of this map consists of those f whose expansions start with z^{-d_i}, i.e., $f \in \mathscr{L}(D)$. On the other hand*, if $\omega \in \Omega_X(-D)$, we can define a map

$$l_\omega : V \longrightarrow \mathbb{C}$$

by

$$l_\omega\left(\left\{\sum_j a_{i,j} z_i^{-j}\right\}\right) = \sum_{i=1}^{n} \operatorname{res}_{x_i}\left[\left(\sum_j a_{i,j} z_i^{-j}\right) \cdot \omega\right]$$

and as above, l_ω is zero on the image of $\mathscr{L}(D_m)$ in V. Moreover, writing out the residue one sees that $l_\omega \equiv 0$ if and only if $\omega \in \Omega_X(-D_m)$. Thus:

$$\dim V \geqq \dim[\Omega_X(-D)/\Omega_X(-D_m)] + \dim[\mathscr{L}(D_m)/\mathscr{L}(D)].$$

We assume m is large enough so that $\deg D_m > \deg K_X$: then no differential ω can have zeroes on D_m so $\Omega_X(-D_m) = (0)$. Moreover, assume m is large enough so that $\dim \mathscr{L}(D_m) = \dim \mathscr{L}(mH) = P_X(m)$ where P_X is the Hilbert polynomial of X. Then one gets

$$\deg(D_m - D) = \sum(e_i - d_i)$$

$$= \dim V$$

$$\geqq \dim \Omega_X(-D) + \dim \mathscr{L}(D_m) - \dim \mathscr{L}(D)$$

$$= \dim \mathscr{L}(K_X - D) + P_X(m) - \dim \mathscr{L}(D).$$

*In case of confusion at this point, let me stress: if, e.g., D is a positive divisor, $\mathscr{L}(D)$ or $\Omega_X(D)$ means poles are allowed on D while $\mathscr{L}(-D)$ or $\Omega_X(-D)$ means zeroes are required on D.

But as in §6C, we have

$$P_X(m) = m \cdot \deg X - p_a(X) + 1$$
$$= \deg D_m - p_a(X) + 1$$

hence

(*) $\dim \mathscr{L}(D) - \dim \mathscr{L}(K_X - D) \geqq \deg D - p_a(X) + 1.$

This formula is valid for any divisor D. Therefore, we may replace D by $K_X - D$ in it and get also:

$$\dim \mathscr{L}(K_X - D) - \dim \mathscr{L}(D) \geqq \deg(K_X - D) - p_a(X) + 1.$$

But we know that $\deg K_X = 2p_a(X) - 2$ (7.9), hence this reduces to exactly (*) with the opposite inequality! This proves the Riemann-Roch formula with $p_a(X)$ as g. But as a final step, take $D = 0$ in this formula and we find:

$$\dim \mathscr{L}(0) - \dim \mathscr{L}(K_X) = -p_a(X) + 1.$$

But $\mathscr{L}(0) =$ functions with no poles, i.e., constants, and $\mathscr{L}(K_X) = \Omega_X$, so this says $1 - p_g(X) = 1 - p_a(X)$ and all parts of (7.26) are proven.   QED

The Riemann-Roch theorem is a marvelous tool in investigating both the function theory and geometry on a smooth curve X. Some of the standard references for this are Hensel-Landsberg [1], Severi [1], Coolidge [1], Serre [1] and a recent survey is contained in Mumford [2]. Here we will merely sketch some of the first consequences for the classification of curves. First look at a low genera:

0) if $g = 0$, then we have seen $X = \mathbb{P}^1$. Note that $|K_X| = \phi$, $\deg K_X = -2$. These curves are called *rational*.

1) if $g = 1$, then $\dim \Omega_X = 1$ and $\deg K_X = 0$. It follows that, up to scalars, X admits exactly one regular differential ω and that this ω has no zeroes either:

$$(\omega) = 0.$$

Moreover, we saw in §7A that X was in biregular correspondence with a smooth plane cubic. Such curves are called *elliptic*. We will study them more closely in §7D.

2) if $g = 2$, then $\dim \Omega_X = 2$ and $\deg K_X = 2$. If ω_1, ω_2 are a basis of Ω_X, then $f = \omega_1/\omega_2$ defines a map

$$\pi : X \longrightarrow \mathbb{P}^1.$$

Then* $\pi^{-1}(0) = (\omega_1), \pi^{-1}(\infty) = (\omega_2)$ and since $\deg K_X = 2$, π has degree 2. According to (6.20), if $B =$ branch points of π, then

$$K_X \equiv \pi^{-1}(K_{\mathbb{P}^1}) + B \equiv \pi^{-1}(-2(0)) + B,$$

hence $\deg B = 6$, i.e., π has 6 branch points. Moreover, $\mathbb{C}(X)$ is a quadratic field extension of $\mathbb{C}(\mathbb{P}^1)$, hence if t is a coordinate on $\mathbb{P}^1$,

---

*It might happen that $(\omega_1), (\omega_2)$ have a common zero x_0, in which case $\pi^{-1}(0) = (\omega_1) - x_0, \pi^{-1}(\infty) = (\omega_2) - x_0$. But then π would have degree 1 and hence be biregular, hence X has, in fact, genus 0 and not 2.

$$\mathbb{C}(X) \cong \mathbb{C}(t, \sqrt{f(t)})$$

for some polynomial f. It is not hard to see the branch points of π are precisely the points of X over the zeroes of f of odd order or the point of X over $t = \infty$ if the degree of f is odd. So changing f to f/g^2 for suitable g, we find that degree $f = 5$ or 6.

In general, curves X of genus ≥ 2 which admit a degree 2 map

$$\pi : X \longrightarrow \mathbb{P}^1$$

or, equivalently, such that $\mathbb{C}(X)$ is a quadratic extension of a purely transcendental subfield $\mathbb{C}(t)$, are called *hyperelliptic*. Using Riemann-Roch, one proves that the other curves X have another explicit form:

(7.27) **Proposition.** *If X is a smooth non-hyperelliptic curve of genus $g \geq 3$, then the linear system $|K_X|$ of canonical divisors (ω) defines a regular map*

$$Z : X \longrightarrow \mathbb{P}^{g-1}$$

which is biregular between X and a smooth curve $X' = Z(X)$ in $\mathbb{P}^{g-1}$ of degree $2g - 2$ whose hyperplane sections are precisely the canonical divisors (ω), $\omega \in \Omega_X$.

Such curves X' are called *canonical curves*. The simplest example is given by smooth quartic plane curves in which case $g = 3$.

The proof is based on:

(7.28) **Lemma.** *$|K_X|$ has no base points, and for all $x \in X$, $|K_X - x|$ has no base points.*

Proof of Lemma. If $|K_X|$ had a base point x, it would mean that every $\omega \in \Omega_X$ vanished at x, hence $\Omega_X = \Omega_X(-x)$. But dim $\Omega_X = g$ and by Riemann-Roch

$$\dim \ \mathscr{L}(x) - \dim \Omega_X(-x) = 1 - g + 1 = 2 - g.$$

So if $\Omega_X = \Omega_X(-x)$, we find dim $\mathscr{L}(x) = 2$, i.e., besides constants, there are other functions $f \in \mathbb{C}(X)$ with only a simple pole at x. Such an f defines a map $\pi : X \to \mathbb{P}^1$ such that $\pi^{-1}(\infty) = x$, hence deg $\pi = 1$, hence X is biregular with $\mathbb{P}^1$, hence $g = 0$: contradicting our hypothesis!

Similarly if $|K_X - x|$ has a base point y, then $\Omega_X(-x) = \Omega_X(-x-y)$. We have just seen that dim $\Omega_X(-x) = g - 1$, and by Riemann-Roch

$$\dim \ \mathscr{L}(x + y) - \dim \Omega_X(-x-y) = 2 - g + 1 = 3 - g.$$

So if $\Omega_X(-x) = \Omega_X(-x-y)$, we find dim $\mathscr{L}(x + y) = 2$. As above there is now a degree 2 map $\pi : X \to \mathbb{P}^1$ and this means X is hyperelliptic: also contradicting our hypothesis.   QED

Going back to the theory of §6A, by the lemma $|K_X|$ defines a map $Z : X \to \mathbb{P}^{g-1}$ such that the divisors $Z^{-1}[H]$ are equal to the divisors (ω), $\omega \in \Omega_X$. By the lemma, for all $x, y \in X$, $x \neq y$, we can find a divisor $D \in |K_X - x|$ with $y \notin$ Support x. Then $D + x = (\omega)$, where ω is zero at x but not at y. If $(\omega) = Z^{-1}[H]$, it follows that $Z(x) \in H$ but $Z(y) \notin H$, hence Z is injective. Thus if $X' = Z(X)$,

$$\begin{aligned}
\deg X' &= \deg(X' \cdot H) \\
&= \deg Z^{-1}[H] \\
&= \deg(\omega) \\
&= 2g - 2.
\end{aligned}$$

Finally, by the lemma, for all x, there is a divisor $D \in |K_X - x|$ with $x \notin \text{Support } D$. Then $D + x = (\omega)$ where ω has a *simple* zero at x. Therefore (ω) has $2g - 3$ zeroes besides x. If $(\omega) = Z^{-1}[H]$, it follows that H meets X' at $Z(x)$ and at $2g - 3$ further points. Thus $\text{mult}_x \cdot Z(x) = 1$. This proves that X' is smooth, hence by (7.1), Z is biregular between X and X'.   QED

§ 7D. Curves of Genus 1 as Plane Cubics and as Complex Tori $\mathbb{C}/L$

Let X be a smooth curve of genus 1, or *elliptic* curve. We have seen in §7A that there is a biregular correspondence between X and a smooth cubic $X^* \subset \mathbb{P}^2$. We have seen in §7B that topologically X is homeomorphic to a torus:

We have seen in §7C that X carries a unique (up to scalars) regular differential form ω without zeroes or poles. In this section I want to try to pull these stands together.

Let me begin by showing that every smooth cubic $X^* \subset \mathbb{P}^2$ has a simple normal form in a suitable coordinate system and that we can then write down the form ω. In fact, according to (7.13), if $X^* = V(f)$, and if $H(f)$ is the Hessian of f:

$$H(f) = \det_{0 \leq i,j \leq 2} \frac{\partial^2 f}{\partial x_i \partial x_j}$$

$X^* \cap V(H(f))$ is the set of points of inflexion of X^*. Note that $\deg H(f) = 3$, so $V(H(f))$ is another cubic curve. Therefore, counted perhaps with some multiplicities, X^* has 9 points of inflexion. Now choose coordinates (X_0, X_1, X_2) for $\mathbb{P}^2$ so that $(0,0,1)$ is a point of inflexion of X^* and $X_0 = 0$ is the tangent line to X^* there. Then the intersection of the cubic X^* and the line $V(X_0)$ must consist only in $(0,0,1)$, counted with multiplicity 3, i.e.,

$$f = X_1^3 + X_0 \cdot (\text{quadratic form}),$$

or

$$\begin{aligned}
-f = {}& a X_0 \cdot X_2^2 - X_0 X_2 (b X_0 + c X_1) \\
& - (X_1^3 + d X_1^2 X_0 + e X_1 X_0^2 + f X_0^3).
\end{aligned}$$

Then $a \neq 0$ since $(0,0,1)$ is a non-singular point of X^* and replacing X_2 by

$$X_2' = (\sqrt{a}\, X_2 - \frac{1}{2\sqrt{a}}(b\, X_0 + c\, X_1)),$$

the cubic takes the form:

$$f' = X_0 X_2'^2 - (X_1^3 + d'X_1^2 X_0 + e'X_1 X_0^2 + f'X_0^3).$$

Finally, replacing X_1 by

$$X_1' = X_1 + \tfrac{1}{3}d'X_0,$$

the cubic now is given by:

$$f'' = X_0 X_2'^2 - (X_1'^3 + e''X_1' X_0^2 + f''X_0^3).$$

Dropping the primes, we find that in suitable coordinates, every smooth cubic is given by

$$X_0 X_2^2 = X_1^3 + a\, X_1 X_0^2 + b\, X_0^3$$

or in affine coordinates $y = X_2/X_0, x = X_1/X_0$:

(7.29) $$y^2 = x^3 + ax + b$$

This is called *Weierstrass normal form*. Conversely, the condition that such an equation define a smooth cubic is simply that the equation $x^3 + ax + b$ have 3 distinct roots. In fact, one checks that the point $(0,0,1)$ at infinity is always smooth on a cubic given by (7.29) and at finite points, (x_0, y_0) is smooth iff

either $$\frac{\partial}{\partial y}(y^2 - x^3 - ax - b) = 2y$$

or $$\frac{\partial}{\partial x}(y^2 - x^3 - ax - b) = -3x^2 - a$$

is non-zero. The only possible singular points are therefore $(x_0, 0)$ where $x_0^3 + ax_0 + b = 3x_0^2 + a = 0$. But this means x_0 is a double root of $x^3 + ax + b$.

Now if X^* is given by (7.29), I claim that the regular form ω is dx/y. In fact, on X^*:

$$2y\, dy = (3x^2 + a)dx,$$

hence

$$\frac{dx}{y} = \frac{2\, dy}{3x^2 + a}.$$

When X^* is smooth, there is no point on X^* where $y = 0$ and $3x^2 + a = 0$. Thus dx/y is a differential with no poles in the affine piece $X^* - V(X_0) \cap X^*$. At the point at ∞, $(0,0,1)$, one can use affine coordinates $u = X_0/X_2 = 1/y$ and $v = X_1/X_2 = x/y$ and we find that X^* has the equation

$$u = v^3 + avu^2 + bu^3.$$

Thus v is a local analytic coordinate on X^* at $(0,0,1)$ and a small calculation shows

$$\frac{dx}{y} = \frac{2}{2auv + 3bu^2 - 1} \cdot dv$$

which has no pole at $(0,0,1)$. Thus $\omega = dx/y$.

The fundamental fact in the theory of elliptic curves is that, by using this differential ω, we can refine the topological homeomorphism of X with a torus to an analytic isomorphism of X with the analytic quotient space $\mathbb{C}/L$, where L is a lattice in $\mathbb{C}$ acting on $\mathbb{C}$ by translation. In fact, let P_0 be any base point on X and consider the indefinite integral

$$E(P) = \int_{P_0}^{P} \omega$$

taken on paths in X. Since ω is a closed 1-form, locally $E(P)$ is a single-valued analytic function of P, independent of the path chosen to connect P_0 to P. In Weierstrass form, E becomes the well-known elliptic integral:

$$E(x) = \int_{\substack{(x_0,y_0) \\ (\text{path in } X\ast)}}^{(x,y)} \frac{dx}{y} = \int_{\substack{x_0 \\ \text{path in double} \\ \text{covering of } x\text{-plane} \\ \text{defined by } \sqrt{x^3+ax+b}}}^{x} \frac{dx}{\sqrt{x^3 + ax + b}}$$

Globally, $E(P)$ however depends on the choice of path and is only well-defined up to addition of a "period" of ω, i.e., up to:

$$\pi(\gamma) = \int_{\gamma} \omega$$

where γ is a 1-cycle on X.

Note that if γ is a boundary, i.e., $\gamma = \partial\sigma$, σ a 2-chain, then

$$\pi(\gamma) = \int_{\partial\sigma} \omega = \int_{\sigma} d\omega = 0.$$

And $\pi(\gamma_1 + \gamma_2) = \pi(\gamma_1) + \pi(\gamma_2)$. Thus π is a homomorphism from the 1st homology group $H_1(X)$ to $\mathbb{C}$(cf. §5C, discussion of DeRham's theorem).

If $\tilde{X}$ is the universal covering space of X, then globally E defines a function

$$\tilde{E} : \tilde{X} \longrightarrow \mathbb{C}$$

or a function

$$E : X \longrightarrow \mathbb{C}/L$$

where L is the group of periods $\{\pi(\gamma) \,|\, \gamma \in H_1(X)\}$.

(7.30) **Lemma.** *L is a lattice in* $\mathbb{C}$.

Proof. We know that the 1st Betti number of a torus is 2, so $H_1(X) \cong \mathbb{Z} \oplus \mathbb{Z}$. Thus if L is not a lattice, L is contained in a real one-dimensional subspace of $\mathbb{C}$, hence for some $\alpha \in \mathbb{C}, \alpha \neq 0$, we would have

$$\mathrm{Re}(\alpha \cdot \pi(\gamma)) = 0, \quad \text{all } \gamma.$$

Then $\mathrm{Re}(\alpha \cdot E)$ would have no periods, hence would be a single-valued harmonic function on X. But as X is compact, it would have a maximum, which is impossible unless $\mathrm{Re}(\alpha \cdot E)$ is a constant. In this case $\alpha \cdot E$ would be a constant and its differential $\alpha \omega$ would be 0, which it is not.   QED

Thus $\mathbb{C}/L$ is an analytic torus and we can state the main result:

(7.31) **Theorem.** *The elliptic integral* $E = \int \omega$ *defines an analytic isomorphism of* X *with* $\mathbb{C}/L$, *where* L *is the lattice of periods of* ω.

Proof. In the first place, $dE = \omega$ is nowhere zero, hence the function E is everywhere a local coordinate on X, i.e., E is a local homeomorphism from X to $\mathbb{C}/L$. Moreover, as X is compact, the map

$$E: X \longrightarrow \mathbb{C}/L$$

must make X into a finite-sheeted covering of $\mathbb{C}/L$. But by the general theory of coverings, such coverings all correspond to subgroups of $\pi_1(\mathbb{C}/L)$, i.e., they are:

$$\mathbb{C}/L' \longrightarrow \mathbb{C}/L$$

where $L' \subset L$ is a sublattice. This means that E factors:

$$X \xrightarrow[\text{homeo,}]{\approx} \mathbb{C}/L' \longrightarrow \mathbb{C}/L.$$

But then its lifting $\tilde{E}: \tilde{X} \to \mathbb{C}$ satisfies $\tilde{E}(x_1) - \tilde{E}(x_2) \in L'$ for all $x_1, x_2 \in \tilde{X}$ over the same point of X, hence all periods $\pi(\gamma)$ lie in L'. Since L is by definition the set $\{\pi(\gamma)\}$, it follows that $L = L'$.   QED

(7.32) **Corollary.** *An elliptic curve* X *can be given a structure of commutative group in which addition* $+ : X \times X \to X$ *and inverse* $- : X \to X$ *are defined by regular correspondences.*

Proof. In fact, $\mathbb{C}$ is a group under $+$ and L is a subgroup so $\mathbb{C}/L$ is a group. Apply the Theorem and (4.14).

We do not want to digress too far into complex analysis but we cannot help mentioning (without proofs) the converse: if $L \subset \mathbb{C}$ is a lattice, then $\mathbb{C}/L$ is analytically isomorphic to a non-singular plane cubic curve. To prove this, clearly one must construct "doubly periodic" meromorphic functions f on $\mathbb{C}$, i.e., such that $f(z + \lambda) = f(z)$, all $\lambda \in L$. The simplest such is the Weierstrass $\wp$-function:

$$\wp(z) = \frac{1}{z^2} + \sum_{\substack{\lambda \in L \\ \lambda \neq 0}} \left[\frac{1}{(z + \lambda)^2} - \frac{1}{\lambda^2} \right].$$

It can be proven that $\wp$ and its derivative set up such an isomorphism as follows:
define

$$\pi : \mathbb{C} \longrightarrow \mathbb{P}^2$$

by

$$\pi(\lambda) = (0, 0, 1) \quad \text{if} \quad \lambda \in L$$

$$\pi(z) = (1, \wp(z), \wp'(z)) \quad \text{if} \quad z \notin L.$$

Then π factors through $\mathbb{C}/L$ and defines an isomorphism of $\mathbb{C}/L$ with a plane cubic curve in Weierstrass form (7.29). For details, see Lang [3].

Looking again at Corollary (7.32), we see that by a long detour through analysis, we have found a very fine correspondence $+$. Surely, if such a correspondence exists, there must be a direct and hopefully beautiful, purely algebraic definition of $+$. This is indeed the case. Take again the plane cubic curve (7.29) and normalize the elliptic integral E by choosing $P_0 = (0, 0, 1)$ as base point:

$$E(P) = \int_{P_0}^{P} \frac{dx}{y}.$$

Then the isomorphism $E : X \to \mathbb{C}/L$ carries $(0, 0, 1)$ to the origin, i.e., to the identity e for the group law and I claim:

(7.33) **Theorem.** *For any 3 points* $Q_1, Q_2, Q_3 \in X$,

$$\begin{bmatrix} \exists \ a \ line \ l \ such \ that \\ l \cdot X = Q_1 + Q_2 + Q_3 \end{bmatrix} \longleftrightarrow \begin{bmatrix} Q_1 + Q_2 + Q_3 = P_0 \ in \ the \\ group \ law \ on \ X \end{bmatrix}.$$

Proof. Note that, using the property on either side of the equivalence, and Q_1 and Q_2 determine one and only one Q_3 for which the property holds: i.e., $Q_3 = -(Q_1 + Q_2)$ for the property on the right; and $Q_3 = $ 3rd point of intersection in $\overline{Q_1 Q_2} \cdot X$ for the property on the left (if $Q_1 = Q_2$, take $\overline{Q_1 Q_2}$ to be the tangent line to X through Q_1). Therefore it suffices to prove either "$\Rightarrow$" or "$\Leftarrow$" and the other direction follows. In fact, we will prove "$\Rightarrow$". Let L be the linear form defining l and consider the rational function L/X_0, and the map

$$\pi : X \longrightarrow \mathbb{P}^1$$

that it defines. In general, degree$(\pi) = 3$, $\pi^{-1}(0) = l \cdot X$, $\pi^{-1}(\infty) = 3P_0$. But it may happen that a zero and pole of L/X_0 cancel, i.e., $P_0 \in l$, in which case degree $(\pi) = 2$, $\pi^{-1}(0) = l \cdot X - P_0$, $\pi^{-1}(\infty) = 2P_0$. Be that as it may, we can construct the following map:

$$\tau : \mathbb{P}^1 \longrightarrow X$$

$$\tau(t) = \begin{bmatrix} \text{sum in the group law on } X \\ \text{of the 3 (or 2) points } \pi^{-1}(t) \end{bmatrix}.$$

This is clearly a well-defined analytic, and hence, by Chow's theorem, algebraic map. But we saw in (7.20) that curves of lower genus never cover curves of higher genus, i.e., $\tau = $ constant. Since

$$\tau(\infty) = \text{sum of pts. in } \pi^{-1}(\infty) = P_0$$

it follows that

$$P_0 = \tau(0) = \text{sum of pts. in } \pi^{-1}(0) = Q_1 + Q_2 + Q_3. \quad \text{QED}$$

(7.34) **Corollary.** *The points of inflexion on X are the points of order 3 in the group law and are mapped by E to $\frac{1}{3}L/L$. Hence, as a group, they are isomorphic to $(\mathbb{Z}/3\mathbb{Z})^2$ and there are 9 of them.*

Proof. $P \in X$ is a point of inflexion if and only if $l \cdot X = 3P$ for some l, hence if and only if $3P = 3P_0$. Since the points of order 3 in $\mathbb{C}/L$ are the points $\frac{1}{3}L/L$ and $L \cong \mathbb{Z}^2$, the second part follows. QED

The left-hand side in the Theorem gives us now a way of defining the group law on X by purely "synthetic" means, i.e., by a plane construction in the sense of High School geometry. First note that if we take $Q_1 = P_0$ at infinity, $Q_2 = (x_0, y_0)$ in affine coordinates, then $\overline{Q_1 Q_2}$ is the vertical line $x = x_0$; hence the third intersection of $\overline{Q_1 Q_2}$ and X is just $Q_3 = (x_0, -y_0)$: i.e., inverse on X is given by:

$$-(x_0, y_0) = (x_0, -y_0).$$

This being the case, the sum $(x_0, y_0) + (x_1, y_1)$ of 2 points on X is found as follows:

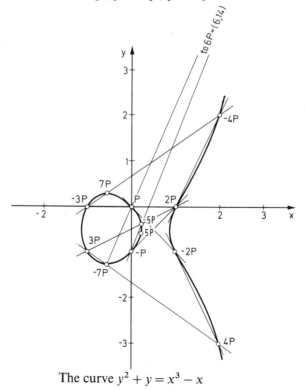

The curve $y^2 + y = x^3 - x$

i) find the 3rd intersection (x_2, y_2) of $\overline{(x_0, y_0), (x_1, y_1)}$ with X

ii) reflect in the x-axis.

Starting from this construction, it is an interesting problem to prove synthetically that X becomes a group, e.g., $+$ is associative! (Cf. Walker [1], p. 191).

There is another more important algebraic interpretation of the group law on X: by the Riemann-Roch theorem on X, one proves that every divisor of degree 1 on X is linearly equivalent to a unique divisor P consisting of one point with multiplicity 1. Therefore every divisor D of degree 0 is linearly equivalent to a unique divisor $P - P_0$. In this way, we get a bijective map:

$$\phi : X \longrightarrow \begin{bmatrix} \text{subgroup of Pic}(X) \text{ of divisors } D \text{ of} \\ \text{degree 0, modulo linear equivalence} \end{bmatrix}$$

$$P \longmapsto (P - P_0) \text{ mod linear equivalence.}$$

The right hand side is a group and in fact, ϕ is an isomorphism of groups. To see this, note that

$$\begin{bmatrix} Q_1 + Q_2 + Q_3 = P_0 \\ \text{in group law on } X \end{bmatrix} \Longrightarrow \exists l, l \cdot X = Q_1 + Q_2 + Q_3$$

$$\Longrightarrow \exists L, (L/X_0) = Q_1 + Q_2 + Q_3 - 3P_0$$

$$\Longrightarrow \begin{bmatrix} \text{the divisor } (Q_1 - P_0) + (Q_2 - P_0) + (Q_3 - P_0) \\ \text{is linearly equivalent to 0} \end{bmatrix}$$

The higher-dimensional generalizations of elliptic curves are called *abelian varieties*: these are by definition non-singular projective varieties with a group structure in which $+$ and $-$ are defined by regular correspondences. They have been much studied: it turns out that all abelian varieties are complex tori, but that conversely, not all complex tori are abelian varieties. See Mumford [1]. It turns out that if X is a curve of genus g, then there is a canonical g-dimensional abelian variety J, called the *Jacobian of X* and an isomorphism

$$\phi : J \xrightarrow{\approx} \begin{bmatrix} \text{subgroup of Pic}(X) \text{ of divisors } D \\ \text{of degree 0, mod linear equivalence} \end{bmatrix}.$$

See Serre [1].

Chapter 8. The Birational Geometry of Surfaces

§ 8A. Generalities on Blowing up Points

We will be interested primarily in non-singular surfaces. Suppose X_1, X_2 are 2 such and $Z \subset X_1 \times X_2$ is a birational map between them. Unlike the case of curves, Z need not be biregular. It follows immediately from the results of Ch. 3, however, that the set

$$F_1 = \{x \in X_1 \mid \dim Z[x] \geq 1\}$$

is 0-dimensional, hence finite and res $Z: X_1 - F_1 \to X_2$ is regular. The points $x \in F_1$ are called the *fundamental points* on X_1 of Z and the components E of their images $Z[x]$ are called the *exceptional curves* on X_2. By symmetry we get a finite set F_2 of fundamental points $x \in X_2$ such that $\dim Z^{-1}[x] \geq 1$, each of which is blown up on X_1 to a finite set of exceptional curves on X_1. The best we can say about Z is that it sets up a biregular correspondence between $X_1 - F_1 - Z^{-1}[F_2]$ and $X_2 - F_2 - Z[F_1]$. A typical example is given by the Cremona transformation $(x,y) \longmapsto \left(\dfrac{1}{x}, \dfrac{1}{y}\right)$ of $\mathbb{P}^2$ (cf. (2.20)).

We wish to get an overall picture of the totality of non-singular surfaces biratonally equivalent to each other. The best way to set this up is to start as in §7A with a field K finitely generated over $\mathbb{C}$ but now of transcendence degree 2 over $\mathbb{C}$. Then by a *smooth model* of K we mean a non-singular surface $X^2 \subset \mathbb{P}^n$ plus an isomorphism α over $\mathbb{C}$ of $\mathbb{C}(X)$ with K. As we mentioned in Chapter 2, it is known that a smooth model always exists. We will bypass this difficult theorem and simply assume that we are dealing with a K for which a smooth model exists. Now given any 2 smooth models $(X_1, \alpha_1), (X_2, \alpha_2)$ of K, the isomorphism:

$$\mathbb{C}(X_1) \xleftarrow{\ \approx\ }_{\alpha_1} K \xrightarrow{\ \approx\ }^{\alpha_2} \mathbb{C}(X_2)$$

induces a birational correspondence between X_1 and X_2. We now introduce a partial ordering in the set of all models by:

(8.1) Definition. *The model* (X_1, α_1) *dominates* (X_2, α_2) *or* $(X_1, \alpha_1) \geq (X_2, \alpha_2)$ *if the induced birational correspondence* $Z: X_1 \to X_2$ *has no fundamental points, hence is regular.*

In particular, if $(X_1, \alpha_1) \geq (X_2, \alpha_2)$ and $(X_1, \alpha_1) \leq (X_2, \alpha_2)$, then the birational correspondence $Z: X_1 \to X_2$ is biregular and we say that the models are *equivalent*:

$$(X_1, \alpha_1) \sim (X_2, \alpha_2).$$

Our first main point is that the operation of blowing up a point x of $X: X \to B_x(X)$, introduces a canonical way of passing from one equivalence class of models (X, α) and a set of corresponding points $x \in X$, to another equivalence class. For this we need to check 2 points:

(8.2) Proposition. a) *If X is a projective variety (of any dimension) and $U \subset X$ is the set of smooth points, then for all $x \in U$, $B_x(X)$ is smooth at all points over U.*

b) *If X_1, X_2 are 2 varieties (of any dimension) and $Z \subset X_1 \times X_2$ is a birational correspondence under which $U_1 \subset X_1$ and $U_2 \subset X_2$ correspond biregularly, then for all smooth points $x_1 \in U_1, x_2 = Z[x_1] \in U_2$, the induced birational correspondence Z' of $B_{x_1}(X_1)$ and $B_{x_2}(X_2)$ is biregular between the inverse images of U_1 and U_2^*.*

Before proving this, we want to go over the definition of $B_x(\mathbb{P}^n)$ again, introducing this time appropriate affine coordinates:

Let $(X_0, \ldots, X_n)$ resp.$(Y_1, \ldots, Y_n)$ be homogeneous coordinates in $\mathbb{P}^n$ (resp. $\mathbb{P}^{n-1}$).

Let $x = (1, 0, \ldots, 0)$ so $p_x(a_0, \ldots, a_n) = (a_1, \ldots, a_n)$.

Then $B_x(\mathbb{P}^n) = $ (locus of zeros in $\mathbb{P}^n \times \mathbb{P}^{n-1}$ of $X_i Y_j - X_j Y_i, 1 \leq i, j \leq n$). We can cover $B_x(\mathbb{P}^n)$ by $2n$ affines:

$$U_i = B_x(\mathbb{P}^n) \cap [\mathbb{P}^n - V(X_i) \times \mathbb{P}^{n-1} - V(Y_i)], \qquad 1 \leq i \leq n$$

$$V_j = B_x(\mathbb{P}^n) \cap [\mathbb{P}^n - V(X_0) \times \mathbb{P}^{n-1} - V(Y_j)], \qquad 1 \leq j \leq n.$$

I) Under $p_1 : B_x(\mathbb{P}^n) \to \mathbb{P}^n, U_i$ goes isomorphically to the affine $\mathbb{P}^n - V(X_i)$ of $\mathbb{P}^n$. With all the U_i's, we cover that part of $B_x(\mathbb{P}^n)$ which is isomorphic to $\mathbb{P}^n - \{x\}$.

II) Affine coordinates in the ambient space $\mathbb{C}^{2n-1}$ containing V_j are

$$Z_1 = \frac{X_1}{X_0}, \ldots, Z_n = \frac{X_n}{X_0}$$

$$W_1 = \frac{Y_1}{Y_j}, \ldots, W_n = \frac{Y_n}{Y_j} \qquad \text{(no } W_j\text{)}.$$

In these coordinates, V_j is defined by the $n-1$ equations:

(*)
$$Z_1 = Z_j W_1, \ldots, Z_n = Z_j W_n$$

i.e.

$$V_j \cong \mathbb{C}^n$$

$$\text{via } x \to (Z_j(x), W_1(x), \ldots, W_n(x)).$$

III) It is easiest to see what this means if we formally solve the equations (*):

(**)
$$W_1 = \frac{Z_i}{Z_j}, \ldots, W_n = \frac{Z_n}{Z_j}$$

---

*This is also true without the assumption that x_1 is smooth. However a direct proof is somewhat messy.

and say simply that $Z_j, \dfrac{Z_1}{Z_j}, \dots, \dfrac{Z_n}{Z_j}$ are coordinates on V_j. Thus in the process of blowing up, the affine space $\mathbb{P}^n - V(X_0)$ with coordinates $(Z_1, \dots, Z_n)$ is replaced by n affine spaces V_j with coordinates $\left(Z_j, \dfrac{Z_1}{Z_j}, \dots, \dfrac{Z_n}{Z_j} \right)$ appropriately glued.

(IV) Let $E \cong \{x\} \times \mathbb{P}^{n-1}$ be the inverse image of x in $B_x(\mathbb{P}^n)$. Since $\{x\} = V(Z_1, \dots, Z_n)$, it follows that:

$$E \cap V_j = V\left(Z_j \cdot \frac{Z_1}{Z_j}, \dots, Z_j, \dots, Z_j \cdot \frac{Z_n}{Z_j} \right)$$

$$= V(Z_j).$$

Thus $E \cap V_j$ has affine coordinates $\dfrac{Z_1}{Z_j}, \dots, \dfrac{Z_n}{Z_j}$. E as a whole is just the projective space with homogeneous coordinates $(Z_1, \dots, Z_n)$ and can be canonically identified with the projective space of 1-dimensional subspaces of $T_{x, \mathbb{P}^n}$.

The case $n = 2$:

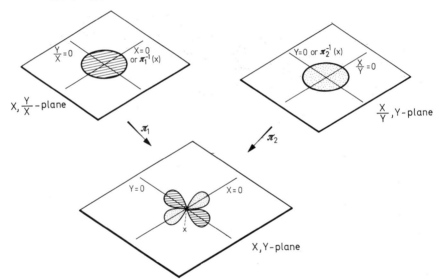

To prove part (a) of the Proposition, note first of all that $B_x(\mathbb{P}^n) - E$ is biregular with $\mathbb{P}^n - \{x\}$, hence $B_x(X) - E_{x,x}$ is biregular with $X - \{x\}$. Therefore $B_x(X) - E_{x,x}$ is smooth at points corresponding to U. The point is to show that $B_x(X)$ is smooth at all points of $E_{x,x}$ too. To work out $B_x(X)$ near these points, use the above coordinates, let $r = \dim X$ and let X be defined near x by $f_1 = \dots = f_{n-r} = 0$, where df_i are independent linear forms at 0. Expand each f_i:

$$f_i(Z_1, \dots, Z_n) = \sum_{j=1}^{n} a_{ij} Z_j + f_i'(Z_1, \dots, Z_n)$$

all terms in f_i' of degree ≥ 2.

Then the functions:

$$f_i\left(Z_k\cdot\frac{Z_1}{Z_k},\ldots,Z_k\cdot\frac{Z_n}{Z_k}\right)=Z_k\cdot\sum_{j=1}^{n}a_{ij}\frac{Z_j}{Z_k}+Z_k^2\cdot f_i''\left(Z_k,\frac{Z_1}{Z_k},\ldots,\frac{Z_n}{Z_k}\right)$$

vanish on $B_x(X)\cap V_k$. Since $B_x(X)$ is by definition the Zariski closure of $X-\{x\}$ in $B_x(\mathbb{P}^n)$, $Z_k\not\equiv 0$ on $B_x(X)\cap V_k$. Therefore

$$f_i^*=\sum_{j=1}^{n}a_{ij}\frac{Z_j}{Z_k}+Z_k\cdot f_i''\left(Z_k,\frac{Z_1}{Z_k},\ldots,\frac{Z_n}{Z_k}\right)$$

vanishes on $B_x(X)\cap V_k$, and therefore the linear functions $\sum_{j=1}^{n}a_{ij}(Z_j/Z_k)$ vanish on $E_{x,X}\cap V_k$. In other words $E_{x,X}$ is contained in the linear subspace L of E with homogeneous equations $\sum_{j=1}^{n}a_{ij}Z_j=0$. Since $x\in X$ is smooth, $\mathrm{rk}(a_{ij})=n-r$, hence $\dim L=r-1$. But $\dim E_{x,X}=r-1$ too, so it follows that $E_{x,X}=L$. Moreover for all $y\in E_{x,X}\cap V_j$, the differentials of the $n-r$ linear functions $\sum_{j=1}^{n}a_{ij}(Z_j/Z_k)$ defining L are independent linear functions on $T_{y,E}$. But these differentials are precisely the restrictions from $T_{y,B_x(\mathbb{P}^n)}$ to $T_{y,E}$ of the differentials df_i^*. Therefore the differentials df_i^* are independent linear functions on $T_{y,B_x(\mathbb{P}^n)}$. Since $\dim B_x(X)=r$, it follows that $B_x(X)$ is smooth at y and is defined locally precisely by the equations $f_1^*=\ldots=f_{n-r}^*=0$.

As a by-product of this argument, since

$$T_{x,X}=\left\{\text{subspace of }T_{x,\mathbb{P}^n}\text{ defined by }\sum_{j=1}^{n}a_{ij}dZ_j=0,\text{ all }i\right\}$$

we get the canonical identification:

$$\left\{\begin{array}{l}\text{Proj. Sp. of 1-dim}\\ \text{subsp. of }T_{x,X}\end{array}\right\}\cong\left\{\begin{array}{l}\text{linear subsp. of }E\\ \text{defined by }\sum_{j}a_{ij}Z_j=0,\text{ all }i\end{array}\right\}=E_{x,X}$$

and since $x\in X$ is smooth, (tangent cone to X at x) = (Zariski tangent space to X at x) and this reproves a result we found in §5A. Recall also from §5A that $B_x(X)$ as a topological space is intrinsically described as the union of an open piece $X-\{x\}$ and a closed piece which is the projective space associated to $T_{x,X}$; reformulating §5A slightly, one finds that the topology may be described like this:

if $y_n\in X-\{x\}$ satisfy $y_n\to x$ as $n\to\infty$, and $D:\mathcal{O}_x\to\mathbb{C}$ is a derivation representing an element of $T_{x,X}$, then

$$\begin{bmatrix}y_n\to\mathbb{C}\cdot D\\ \text{in }B_x(X)\\ (\text{or in }B_x(\mathbb{P}^n))\\ \text{as }n\to\infty\end{bmatrix}\Leftrightarrow\begin{bmatrix}\exists\,\epsilon_n\in\mathbb{C},\epsilon_n\to 0\text{ such that for all}\\ \forall f\in\mathcal{O}_{x,X},Df=\lim_{n\to\infty}\dfrac{f(y_n)-f(x)}{\epsilon_n}\end{bmatrix}$$

Part (b) of the Proposition is a Corollary of this: in the notation of (b), note that the differential $dZ(x_1): T_{x_1,X_1} \overset{\approx}{\longrightarrow} T_{x_2,X_2}$ induces an isomorphism $\overline{dZ}$ of E_{x_1,X_1} with E_{x_2,X_2} and it follows from the above description of the topology on $B_{x_i}(X_i)$ that

$$\text{if } y_n \in X_1 - \{x_1\}, \text{ where } y_n \to x \text{ as } n \to \infty, \text{ then}$$

$$(y_n \text{ has a limit in } B_{x_1}(X_1)) \text{ iff}(Z[y_n] \text{ has a limit in } B_{x_2}(X_2))$$

$$\text{and if these limits exist, they correspond under } \overline{dZ}.$$

Therefore Z' is at least a bijective map between the inverse images of U_i on $B_{x_i}(X_i)$. Then by (a) and Zariski's Main Theorem, it is biregular.

Before leaving these generalities, one more useful remark:

(8.3) **Proposition.** *Let* $x \in X^r \subset \mathbb{P}^n$ *be a smooth point and choose a projection* $p_L : X^r \longrightarrow \mathbb{P}^r$ *which is smooth at* x. *Then* p_L *lifts to a regular correspondence:*

$$B_x(X) \longrightarrow B_{p(x)}(\mathbb{P}^r)$$

which is smooth at all points of $E_{x,X}$.

This is easy to check. In particular if $Z_1,\ldots,Z_r$ are affine coordinates on $\mathbb{P}^r$ near $p(x)$, then $Z_1,\ldots,Z_r$ are analytic coordinates on X in a small classical neighborhood of x, and it follows from that at each point y of $B_x(X)$ over x, for some j,

$$Z_j, \frac{Z_1}{Z_j}, \ldots, \frac{Z_r}{Z_j} \text{ are analytic coordinates in a neighborhood of } y.$$

§8B. Resolution of Singularities of Curves on a Smooth Surface by Blowing up the Surface; Examples

An indispensable tool in the study of the birational geometry of non-singular surfaces is a stronger resolution theorem for singular curves. We shall prove:

(8.4) **Theorem.** *Let* X *be a smooth projective surface and let* $C \subset X$ *be a pure 1-dimensional closed algebraic set (possibly reducible but no component a point). There exists a sequence of blow-ups:*

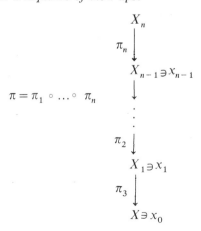

$$\pi = \pi_1 \circ \ldots \circ \pi_n$$

where $X_{i+1} = B_{x_i}(X_i)$ and $\pi_{i+1} : X_{i+1} \to X_i$ is the natural projection such that if

$$F = \{x_0, \pi_1 x_1, \pi_1(\pi_2 x_2), \ldots\ldots, \pi_1(\ldots(\pi_{n-1} x_{n-1})\ldots)\}$$

is the fundamental set of the correspondence π^{-1}, and

$$\tilde{C} = Zariski\ closure\ of\ \pi^{-1}(C - C \cap F)$$

is the so-called proper transform of C on X_n, then $\tilde{C}$ is smooth.

The following proof is due to Hironaka and is extracted from his proof of resolution in the general case (Hironaka[1]). We observe first that the problem is a local one. In fact C has only a finite number of singular points $\{y_1, \ldots, y_t\}$. If we can blow up X at y_1 and at points lying over y_1 on the resulting surfaces until you get an X_{n_1} over X on which the proper transform $\tilde{C}_1$ of C is smooth at all points over y_1, then $\tilde{C}_1$ has singularities only at the points $\{y_2', \ldots, y_t'\}$ corresponding biregularly to $\{y_2, \ldots, y_t\}$. So by induction on t we can eventually resolve C. Fix therefore one singular point y of C and let Z_1, Z_2 be analytic coordinates on X in a neighborhood of y. Since $\mathcal{O}_{y,X}$ is a UFD, the ideal $I(C)_y$ of all $f \in \mathcal{O}_{y,X}$ such that $f \equiv 0$ on C is principle. Any function $f \in \mathcal{O}_{y,C}$ whose divisor (f) equals $\sum C_i$ near y (C_i the components of C) is a generator of $I(C)_y$: we call such an f a *local equation* of C at y. Our first step is to define 2 invariants to measure how bad a singularity C has at y.

(8.5) **Multiplicity.** let f be a local equation of C at y and expand f:

$$f = f_\mu(Z_1, Z_2) + f_{\mu+1}(Z_1, Z_2) + \ldots\ldots$$

f_k homogeneous of degree k

$$f_\mu \not\equiv 0.$$

Then $\mu = \mu_y(C)$ is multiplicity of C at y. More intrinsically, this means that $f \in \mathfrak{M}_{y,X}^\mu$ but $f \notin \mathfrak{M}_{y,X}^{\mu+1}$, and since any 2 f's differ by a unit in $\mathcal{O}_{y,X}$ this μ is independent of the choice of f. More generally any analytic function f can be used to define μ provided that f vanishes to 1st order on each component of C and vanishes nowhere else.

(8.6) **Invariant v.** Let μ = multiplicity of f and after a linear change of coordinates, assume $f_\mu(1,0) \neq 0$. Then by the Weierstrass preparation theorem, we can write:

$$f = (\text{unit}) \cdot (Z_1^\mu + a_1(Z_2) Z_1^{\mu-1} + \ldots\ldots + a_\mu(Z_2))$$

where $a_i(Z_2)$ are convergent power series in Z_2 whose multiplicity (= degree of lowest order terms) is $\geq i$. Next let $\tilde{Z}_1 = Z_1 + \dfrac{a_1(Z_2)}{\mu}$. In terms of the new coordinates $\tilde{Z}_1, Z_2$ one sees immediately that

$$f = (\text{unit}) \cdot (\tilde{Z}_1^\mu + b_2(Z_2) \tilde{Z}_1^{\mu-2} + \ldots + b_\mu(Z_2))$$

i.e., the coefficient of $Z_1^{\mu-1}$ is 0! Also mult $b_i \geq i$ still. Set

$$v = \min_{2 \leq i \leq \mu} \left\lceil \frac{\text{mult}(b_i(Z_2))}{i} \right\rceil.$$

Then $v \in \dfrac{1}{\mu!} \mathbb{Z}$, and $v \geq 1$. This v is in fact independent of the analytic coordinate system chosen, but rather than stopping to check this, we can just as well define $v_y(C)$ to be the minimum of over all coordinate systems Z_1, Z_2 of the above v.

Now let

$$X_1 = B_y(X)$$

$$\pi_1 : X_1 \to X \text{ the projection}$$

$$C_1 = \text{Zar. closure of } \pi_1^{-1}(C - \{x\})$$

$$y(1), \ldots, y(k) = C_1 \cap E_{y,x} = \text{proj. tangent cone of } C.$$

The main step is the assertion:

a) if $v_x(C) = 1$, then $k > 1$ and for all i

$$\mu_{y(i)}(C_1) < \mu_y(C).$$

b) if $v_x(C) > 1$, then $k = 1$ and

either $\mu_{y(1)}(C_1) < \mu_y(C)$

or $\mu_{y(1)}(C_1) = \mu_y(C)$ but $v_{y(1)}(C_1) \leq v_y(C) - 1$.

By an evident induction, this will prove (8.4).

Proof of (a) and (b). Let C be defined locally by

$$f = f_\mu + f_{\mu+1} + \ldots = 0.$$

At all points of $E_{y,x}$, one has local coordinates either $Z_1, \dfrac{Z_2}{Z_1}$ or $Z_2, \dfrac{Z_1}{Z_2}$. In these coordinates, f becomes

$$f = f_\mu\left(Z_1, Z_1 \cdot \frac{Z_2}{Z_1}\right) + f_{\mu+1}\left(Z_1, Z_1 \cdot \frac{Z_2}{Z_1}\right) + \ldots$$

$$= Z_1^\mu f_\mu\left(1, \frac{Z_2}{Z_1}\right) + Z_1^{\mu+1} f_{\mu+1}\left(1, \frac{Z_2}{Z_1}\right) + \ldots$$

$$= Z_1^\mu \underbrace{\left[f_\mu\left(1, \frac{Z_2}{Z_1}\right) + Z_1 \cdot f_{\mu+1}\left(1, \frac{Z_2}{Z_1}\right) + \ldots \right]}_{\text{call this } f_1^*.}$$

or

$$f = Z_2^\mu \underbrace{\left[f_\mu\left(\frac{Z_1}{Z_2}, 1\right) + Z_2 f_{\mu+1}\left(\frac{Z_1}{Z_2}, 1\right) + \ldots \right]}_{\text{call this } f_2^*}$$

At points of the 1st type, note that $V(Z_1)$ is $E_{y,x}$, and since by definition $E_{y,x}$ is not a component of C_1, f_1^* must vanish on C_1. Since f and hence f_1^* vanishes to 1st order on all components of C_1 (since almost everywhere on each of these, X_1 is biregular to X), f_1^* is a local equation of C_1. At points of the 2nd type, similarly f_2^* is a local equation of C_1. In particular, $C_1 \cap E_{y,x}$ is defined by

$$\text{either } Z_1 = f_\mu\left(1, \frac{Z_2}{Z_1}\right) = 0$$

$$\text{or } Z_2 = f_\mu\left(\frac{Z_1}{Z_2}, 1\right) = 0.$$

Identifying $E_{y,x}$ with $\mathbb{P}^1$ with homogeneous coordinates $Z_1, Z_2, C_1 \cap E_{y,x}$ corresponds to the roots of the homogeneous equation $f_\mu(Z_1, Z_2) = 0$. Therefore

$$k = 1 \Leftrightarrow f_\mu \text{ is a } \mu\text{th power of a linear form.}$$

Now let Z_1, Z_2 be coordinates in which

(*) $$f = (\text{unit})[Z_1^\mu + b_2(Z_2)Z_1^{\mu-2} + \ldots + b_\mu(Z_2)],$$

and $v_y(C) = \min\left(\dfrac{\text{mult } b_i}{i}\right)$. Then

$$f_\mu = (\text{cnst.}). \left[Z_1^\mu + \binom{\text{degree } 2}{\text{terms of } b_2} \cdot Z_1^{\mu-2} + \ldots + \binom{\text{degree } \mu}{\text{terms of } b_\mu}\right]$$

hence

$$(f_\mu \text{ is a } \mu\text{th power}) \Leftrightarrow (\text{for all } i, b_i \text{ starts with terms of degree} > i).$$

$$\Leftrightarrow v_y(C) > 1$$

This proves the 1st part of (a) and (b). Next if $v = 1$ let's say $y(i) = \Big($ the point $Z_1 = 0, \dfrac{Z_2}{Z_1} = \alpha\Big)$. Since $k > 1, \dfrac{Z_2}{Z_1} = \alpha$ is a root of $f_\mu\left(1, \dfrac{Z_2}{Z_1}\right) = 0$ of some multiplicity $\mu' < \mu$ and we can write

$$f_1^* = \left(\frac{Z_2}{Z_1} - \alpha\right)^{\mu'} \cdot A + Z_1 \cdot B, \qquad A(y) \neq 0.$$

It follows that if f_1^* is expanded as a power series in the coordinates $\dfrac{Z_2}{Z_1} - \alpha$ and Z_1 near y, then lowest degree terms have order at most μ', i.e., $\mu_{y(i)}(C_1) \leq \mu' < \mu$. This proves the rest of (a). Now say $v > 1$. Then $f_\mu = \text{const. } Z_1^\mu$ and the only point $y(1)$ over y on C_1 is the point $Z_2 = \dfrac{Z_1}{Z_2} = 0$. At this point, using equation (*), we find:

(**) $$f_2^* = (\text{unit}) \cdot \left[\left(\frac{Z_1}{Z_2}\right)^\mu + \frac{b_2(Z_2)}{Z_2^2} \cdot \left(\frac{Z_1}{Z_2}\right)^{\mu-2} + \ldots + \frac{b_\mu(Z_2)}{Z_2^\mu}\right].$$

If $\mu_{y(1)}(C_1) < \mu$ we are through, so assume $\mu_{y(1)}(C_1) = \mu$. Then in equation (**), f_2^*

has been prepared in the coordinates $Z_2, \dfrac{Z_1}{Z_1}$ in the way required to define v, i.e.,

$$f_2^* = (\text{unit})\left[\left(\frac{Z_1}{Z_2}\right)^\mu + b_2'(Z_2)\left(\frac{Z_1}{Z_2}\right)^{n-2} + \ldots + b_\mu'(Z_2)\right]$$

where $b_i'(Z_2) = b_i(Z_2)/Z_2^i$. Therefore

$$v_{y(1)}(C_1) \leq \min_{2 \leq i \leq \mu}\left[\frac{\text{mult } b_i'(Z_2)}{i}\right]$$

$$= \min_{2 \leq i \leq \mu}\left[\frac{\text{mult } b_i(Z_2) - i}{i}\right]$$

$$= v_y(C) - 1.$$

This proves the 2nd part of (b), and hence all of (8.4).

Let's give a few examples to illustrate how this process works:

(A) Say $y \in C$ is a smooth point. Then $\mu = 1$ and the tangent cone $E_{y,C}$ consists in a single point $y^{(1)}$, that corresponding to the subspace $T_{y,C} \subset T_{y,X}$. Since μ can only decrease, $\mu_{y(1)}(C_1) = 1$ too, i.e., $y^{(1)}$ is still smooth on C_1. Moreover C_1 and $E_{y,X}$ cross transversely on $B_y(X)$: in fact if C has local equation

$$f = Z_1 + f_2 + f_3 + \ldots$$

then $y^{(1)}$ is the point $Z_2 = \dfrac{Z_1}{Z_2} = 0$ and at this point $E_{y,X}$ is $Z_2 = 0$ and C_1 has the equation:

$$f_2^* = \frac{Z_1}{Z_2} + Z_2 f_2\left(\frac{Z_1}{Z_2}, 1\right) + \ldots.$$

$$= \frac{Z_1}{Z_2} + f_2(0,1)\cdot Z_2 + \text{higher order terms}.$$

Thus the tangent lines to $E_{y,X}$ and C_1 at $y^{(1)}$ are distinct (note also the tangent to C_1 is determined by the coefficient of Z_2^2 in f, i.e., the 2nd order jet of C).

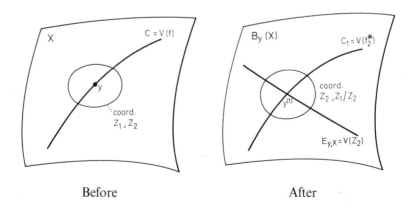

Before   After

(B) Say $y \in C$ is an ordinary double point. Then $\mu = 2$ and the tangent cone $E_{y,C}$ consists in 2 points $y^{(1)}, y^{(2)}$ so by (b) the invariant v must equal 1. In fact C has local equation

$$f = Z_1^2 - Z_2^2 + f_3 + f_4 + \cdots$$

in suitable coordinates. Applying Weierstrass Preparation:

$$f = (\text{unit}) \cdot [\underbrace{Z_1^2 + a_1(Z_2)Z_1 + a_2(Z_2)}_{\text{mult.} \geq 2 \ - Z_2^2 + \text{ higher order terms}}]$$

$$= (\text{unit}) \cdot [\underbrace{\tilde{Z}_1^2 + b_2(Z_2)}_{- Z_2^2 + \beta_3 Z_2^3 + \beta_4 Z_2^4 + \cdots}]$$

In the coordinates $\tilde{Z}_1, Z_2$

$$E_{y,C} = \text{the 2 points } Z_2 = 0, \underbrace{\frac{\tilde{Z}_1}{Z_2} = +1}_{\text{call this } y^{(1)}} \text{ and } Z_2 = 0, \underbrace{\frac{\tilde{Z}_1}{Z_2} = -1}_{\text{call this } y^{(2)}}$$

$$C_1 = \left[\text{ the curve } \left(\frac{\tilde{Z}_1}{Z_2} + 1\right) \cdot \left(\frac{\tilde{Z}_1}{Z_2} - 1\right) + \beta_3 Z_2 + \beta_4 Z_2^2 + \cdots = 0 \right]$$

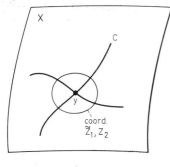

Before

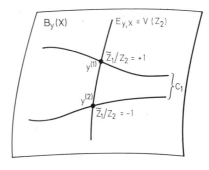

After

In particular, C_1 is smooth at $y^{(1)}$ and $y^{(2)}$ crosses $E_{y,x}$ transversely at 2 points.

(C) Look at double points $y \in C$ with only one tangent line. Their local equations come out

$$f = (\text{unit})[\tilde{Z}_1^2 + b_2(Z_2)]$$

where mult $b_2 \geq 3$. Let $k = \text{mult } b_2 = 2v$. Then $b_2(Z_2) = \eta(Z_2) \cdot Z_v^k, \eta(0) \neq 0$. Then $\eta(Z_2) = \zeta(Z_2)^k$ for some analytic function ζ, hence if $\tilde{Z}_2 = \zeta(Z_2) \cdot Z_2, \tilde{Z}_1$ and $\tilde{Z}_2$ are local coordinates in which f is

$$f = (\text{unit})[\tilde{Z}_1^2 + \tilde{Z}_2^k].$$

Thus up to analytic isomorphism we have only one type of double point for each k. The simplest case is $k = 3$, the *cusp* (cf.§1B, Ex. b). Here

$$E_{y,C} = \left(\text{one point } y^{(1)} \text{ given by } \tilde{Z}_2 = \frac{\tilde{Z}_1}{\tilde{Z}_2} = 0 \right) \text{ and}$$

$$C_1 = \left(\text{the curve} \left(\frac{\tilde{Z}_1}{\tilde{Z}_2} \right)^2 + \tilde{Z}_2 = 0 \right).$$

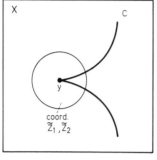

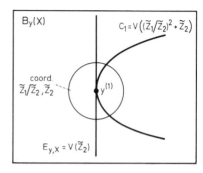

Before	After

We can draw a useful conclusion from this discussion:

(8.7) Corollary. *In any sequence of blow-ups $\pi : X_n \to X$ as in (8.4), for every fundamental point $y \in X$, $\pi^{-1}(y)$ is a union of smooth rational curves, crossing only 2 at a time and then transversely.*

Proof. Use induction on n. If $C = \sum E_i$ is a set of curves on X_{n-1} with this property, then 3 things can happen to $\pi_n^{-1}(C)$: if $x_{n-1} \notin C$, then $\pi_n^{-1}(C) \cong C$; if x_{n-1} is a smooth point of C, say on E_{i_0}, then by Example A above, $\pi_n^{-1}(C)$ has one more component than C, namely $E_{x_{n-1}, X_{n-1}}$, in biregular correspondence with $\mathbb{P}^1$, it meets only E_{i_0} and then transversely; if $x_{n-1} \in E_{i_0} \cap E_{i_1}$, then by Example B above $\pi_n^{-1}(C)$ looks like:

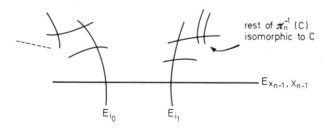

QED

(8.8) Corollary. *Let*

$$Z : X^2 \longrightarrow \mathbb{P}^m$$

be a rational map from a smooth surface X to $\mathbb{P}^m$. Then there is a sequence of blow-ups:

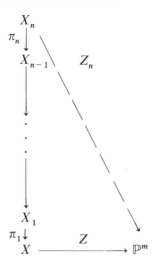

as in (8.4) *such that the induced rational map* $Z_n : X_n \longrightarrow \mathbb{P}^m$ *is everywhere regular.*

Proof. Let Z be defined by the linear system L of divisors on X. Recall that L is the set of divisors $Z^*(H), H \subset \mathbb{P}^m$ a hyperplane, hence L is isomorphic to the set of hyperplanes in $\mathbb{P}^m$ which is also a projective space of dimension m. Choose $2m + 2$ points in L in general position, i.e., no more than m of them in any hyperplane of L, and let these points define divisors $D^{(i)} = Z^*(H^{(i)}), 1 \le i \le 2m + 2$. Let C be the union of all components of all $D^{(i)}$. Apply (8.4) to C, i.e., blow up X until the proper transform $\tilde{C}$ of C on X_n is smooth. I claim that Z_n is regular.

Suppose the linear system L_n defines Z_n and let $D_n^{(i)} = Z_n^*(H^{(i)})$. Let F be the set of fundamental points of π^{-1} and let $E_1, \dots, E_n$ be the exceptional curves, i.e., the components of $\pi^{-1}(F)$. We can decompose $D_n^{(i)}$:

$$D_n^{(i)} = \underbrace{\mathscr{E}^{(i)}}_{\substack{\text{a linear comb.} \\ \text{of } E_j\text{'s}}} + \underbrace{\text{Closure}\left[\pi^{-1}(D^{(i)} - F \cap D^{(i)})\right]}_{\text{proper transform } \tilde{D}^{(i)}}$$

Since $\tilde{C} = \cup \tilde{D}^{(i)}$ is smooth, it follows that the $\tilde{D}^{(i)}$ are all disjoint. Since the bases points of L_n are the fundamental points of Z_n, it suffices to show that for all $y \in X_n$, for some $i, y \notin D_n^{(i)}$. But y is on at most one $\tilde{D}^{(i)}$ and by (8.7) y is on at most two E_j's: say E_{j_1}, E_{j_2}. For all j, let

$$L_{n,j} = \{\text{divisors } D \in L_n \,|\, E_j \text{ is a component of } D\}.$$

Then $L_{n,j}$ is a linear subspace of L_n and since the base points of L_n have codimension $\ge 2, L_{n,j} \subsetneq L_n$. It follows that at most m of the divisors $D_n^{(i)}$ are in each $L_{n,j}$. Therefore at most $2m$ of the divisors $D_n^{(i)}$ have either E_{j_1} or E_{j_2} as a component. But if neither E_{j_1} nor E_{j_2} is a component of $D_n^{(i)}$ and $y \notin \tilde{D}^{(i)}$, then $y \notin D_n^{(i)}$! Since we have $2m + 2$ $D_n^{(i)}$'s, at least one has all these properties. Thus y is not a base point of L_n.

QED

§8C. Factorization of Birational Maps between Smooth Surfaces; the Trees of Infinitely Near Points

We are now ready to try to understand the structure of the partially ordered set of models of a function field K. (8.8) shows us that if $(X_1,\alpha_1),(X_2,\alpha_2)$ are any 2 models, then after a finite number of blow ups, the model obtained from (X_1,α_1) will dominate (X_2,α_2). This is half the story. The 2nd half is contained in:

(8.9) **Theorem.** Let $Z:X_1 \to X_2$ be a birational map between projective surfaces. Let $x_i \in X_i$ be smooth points such that Z is regular at x_1 and $Z[x_1]=x_2$. Then

  either 1) Z^{-1} is regular at x_2, i.e., Z is biregular at x_1

   or 2) the induced birational map W:

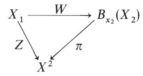

   is regular at x_1.

Proof. Write E for E_{x_2,X_2} and B for $B_{x_2}(X_2)$. Now

$$W[x_1] \subset \pi^{-1}Z[x_1]$$
$$= \pi^{-1}(x_2)$$
$$= E.$$

By Zariski's Main theorem, either $W[x_1]=$ one point and W is regular at x_1, or $W[x_1]=E$. Assume the second holds. Then we want to prove that Z^{-1} is regular at x_2. Now there is an open dense set $E_0 \subset E$ where W^{-1} is regular. Since $x_1 \in W^{-1}[y]$, all $y \in E$, it follows that $W^{-1}[E_0]=\{x_1\}$. For all $y \in E_0$, we get therefore regular maps

$$\begin{bmatrix} \text{neigh of } y \\ \text{in } B \end{bmatrix} \xrightarrow{\;W^{-1}\;} \begin{bmatrix} \text{neigh of } x_1 \\ \text{in } X_1 \end{bmatrix} \xrightarrow{\;Z\;} \begin{bmatrix} \text{neigh of } x_2 \\ \text{in } X_2 \end{bmatrix}$$

hence

(*) $T_{y,B} \xrightarrow{\;dW^{-1}(y)\;} T_{x_1,X_1} \xrightarrow{\;dZ(x_1)\;} T_{x_2,X_2}.$

Since $Z \circ W^{-1}=\pi$, the composition here is $d\pi(y)$. But we can easily compute $d\pi(y)$: in fact let Z_1,Z_2 be analytic coordinates at x_2 in X_2. If $y \in E$ corresponds to the ratio $Z_1 : Z_2 = 1 : \alpha$, $W_1=Z_1, W_2=\dfrac{Z_2}{Z_1}-\alpha$ are analytic coordinates on B with y as origin. Thus locally we have the map

$$Z_1 = W_1$$
$$Z_2 = W_1(W_2+\alpha),$$

whose differential is

$$dZ_1 = dW_1$$

$$dZ_2 = (W_2 + \alpha)dW_1 + W_1 \cdot dW_2.$$

Therefore $d\pi(y)$ is given by the matrix

$$\begin{pmatrix} 1 & 0 \\ \alpha & 0 \end{pmatrix}$$

hence $d\pi(y)$ has rank 1 and the image is a one-dimensional subspace which varies as y varies on E. Therefore by (*)

$$\text{Im}[dZ(x_1)] \supset \sum_{y \in E_0} \text{Im}[d\pi(y)]$$

and, as E_0 has more than 1 point in it, the right hand side is T_{x_2, x_2}. In other words $dZ(x_1)$ is surjective hence injective. This implies that Z^{-1} is regular at x_2—because otherwise $Z^{-1}[x_2]$ would contain a whole curve C through x_1 and $T_{x_1, C}$ would be in the kernel of $dZ(x_1)$. QED

(8.10) **Corollary.** *Let $Z: Y \to X$ be a regular birational map between smooth projective surfaces. Then Z can be factored*:

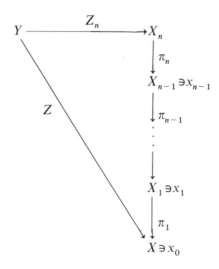

where $X_i = B_{x_{i-1}}(X_{i-1}), \pi_{i+1}: X_{i+1} \longrightarrow X_i$ is the natural projection and Z_n is biregular.

Proof. If Z is not biregular, let x_0 be a fundamental point of Z^{-1}. Then by (8.9), Z factors as $\pi_1 \circ Z_1, Z_1: Y \longrightarrow X_1$ being regular. Continue in this way. Eventually $Z_n: Y \longrightarrow X_n$ must be biregular because each $Z_i: Y \longrightarrow X_i$ has one fewer exceptional curve.

QED

(8.11) **Corollary.** (Natation as in §8.A). *Any 2 smooth models* (X_1,α_1), (X_2,α_2) *of K have a least upper bound* (X_3,α_3),(X_3 *smooth*) *which may be gotten by blowing up suitable points on* X_1 *or on* X_2:

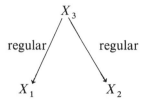

Proof. Let $Z : X_1 \longrightarrow X_2$ be the given birational map. By (8.9), any smooth X^* dominating birationally X_1 and X_2 also dominates the surface X_1' gotten by blowing up one by one the fundamental points $F \subset X_1$ of Z. Let $Z' : X_1' \longrightarrow X_2$ be the new birational map. Then X^* still dominates the surface X_1'' gotten by blowing up one by one the fundamental points $F' \subset X_1'$ of Z'. Continuing like this, we reach an $X_1^{(n)}$ which maps regularly to X_2 such that for any X^*, $X^* \longrightarrow X_1^{(n)}$ is regular, i.e., a least upper bound.   QED

The partially ordered set of models of K has a local analog which is very pretty, called the set of infinitely near points:

(8.12)   i) Fix a projective surface X and a smooth point $x \in X$.
         ii) Look at triples (X^*,x^*,Z): X^* a projective surface, $x^* \in X^*$ a smooth point, $Z : X^* \to X$ a birational map regular at x^* with $Z[x^*] = x$.
         iii) Say

$$(X_1^*,x_1^*,Z_1) \geqq (X_2^*,x_2^*,Z_2)$$

if the induced birational map

$$Z_{12} : X_1^* \longrightarrow X_2^*$$

is regular at x_1^* and $Z_{12}[x_1^*] = x_2^*$.
      iv) Say $(X_1^*,x_1^*,Z_1) \sim (X_2^*,x_2^*,Z_2)$ if

$$(X_1^*,x_1^*,Z_1) \geqq (X_2^*,x_2^*,Z_2) \text{ and } (X_1^*,x_1^*,Z_1) \leqq (X_2^*,x_2^*,Z_2).$$

[Note that equivalence implies that $x_i^* \in X_i^*$ has a small classical neighborhood U_i^* such that $\exists$ a homeomorphism h_{12} fitting into a diagram:

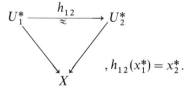

$, h_{12}(x_1^*) = x_2^*.$

The reader can check as an exercise the converse: 2 triples are equivalent if U_i^* and h_{12} exist. This shows that equivalence classes are characterized by the germ of the topological map

$$\text{res } Z : \left(\begin{array}{c}\text{small neigh}\\ \text{of } x^* \in X^*\end{array}\right) \longrightarrow X. \Big]$$

v) An equivalence class of triples we call an *infinitely near point* of X over x. From (8.9) we deduce easily:

(8.13) Proposition. *Every infinitely near point of X over x is represented by a unique triple (X_n, x_n, π), where X_n is gotten by a sequence of blow ups:*

$$X_n \xrightarrow[\pi_n]{} X_{n-1} \xrightarrow[\pi_{n-1}]{} \cdots \cdots \longrightarrow X_1 \xrightarrow[\pi_1]{} X$$

$$\begin{array}{ccc} \omega & & \omega \qquad \omega \\ X_{n-1} & & X_1 \qquad x = x_0 \end{array}$$

where

$$X_{i+1} = B_{x_i}(X_i),$$

$$\pi = \pi_1 \circ \cdots \circ \pi_n,$$

$$x_{i-1} = \pi_i(x_i), 1 \leq i \leq n.$$

(Proof left to reader.)

(8.14) Corollary. *The set of infinitely near points forms a tree:*

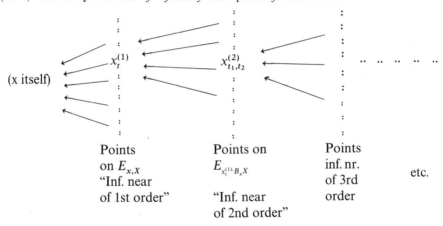

| Points on $E_{x,X}$ "Inf. near of 1st order" | Points on $E_{x_t^{(1)}, B_x X}$ "Inf. near of 2nd order" | Points inf. nr. of 3rd order | etc. |

To whet the reader's appetite, let me mention 2 beautiful results that can be proven with these concepts:

(8.15) If $C \subset \mathbb{P}^2$ is a plane curve of order n, and C' is a smooth curve birational to C, then

$$p_a(C') = \frac{(n-1)(n-2)}{2} - \sum_{\substack{\text{all pts. } P \\ \text{ordinary and} \\ \text{inf. near of } \mathbb{P}^2}} \frac{\mu_P(\mu_P - 1)}{2}$$

where μ_P = multiplicity at P of C (or of its proper transform on the surface containing P).

(This generalizes the formula in Step IIa of the proof of (7.9); cf. Van der Waerden [1]).

(8.16) If K is a function field of transcendence degree 2, then the set of smooth models of K has a unique *minimal* element in it unless $K \cong K_0(t)$, where tr.d$_c K_0 = 1$, i.e., unless K has a model $C \times \mathbb{P}^1$, C a curve. Surfaces birational to $C \times \mathbb{P}^1$ are called *ruled surfaces*. See Šafarevič [1].

§8D. The Birational Map between $\mathbb{P}^2$ and the Quadric and Cubic Surfaces; the 27 Lines on a Cubic Surface

We want to illustrate the preceding ideas with the smooth surfaces $X \subset \mathbb{P}^3$ of degrees 2 and 3. First take X to be a quadric. As we saw in §5A, projecting X from a point $x \in X$ defines a birational map from X to $\mathbb{P}^2$; more precisely, we get a diagram

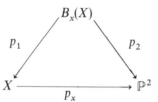

where p_2 is a regular birational map. A curve C on $B_x(X)$ is mapped by p_2 to a point if and only if $p_1(C)$ is a line on X through x. But there are exactly 2 lines $l_1, l_2 \subset X$ through x, and their proper transforms $\tilde{l}_1, \tilde{l}_2 \subset B_x(X)$ are the exceptional curves of p_2. Let $p_x(l_i - \{x\}) = p_2(\tilde{l}_i) = P_i$. Then by (8.9), p_2 factors through $B_{P_1}(B_{P_2}(\mathbb{P}^2))$. Since there are no further exceptional curves on $B_x(X)$, we get a diagram:

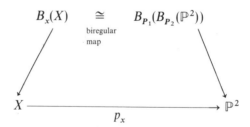

Finally, what linear system L on $\mathbb{P}^2$ defines the rational map $p_x^{-1} : \mathbb{P}^2 \to X \hookrightarrow \mathbb{P}^3$? For all planes $H \subset \mathbb{P}^3$ such that $x \notin H$, $p_x(X \cap H)$ is a conic in $\mathbb{P}^2$, so L is a linear system of conics. Also, since p_1 and p_2 are fundamental points of p_x^{-1}, the conics in L must contain p_1 and p_2. Now the set of all conics is a $\mathbb{P}^5$ and the set of those containing 2 fixed distinct points forms a 3-dimensional subspace. So

$$L = \text{(the set of all conics in } \mathbb{P}^2 \text{ containing } p_1 \text{ and } p_2).$$

Note that 2 conics $C_1, C_2 \in L$ meet in general in 2 points of $\mathbb{P}^2$ besides P_1, P_2; these points correspond under the birational map p_x^{-1} to the 2 points of X in which the 2 hyperplane sections $X \cap H_1, X \cap H_2$ meet (where $C_i = p_x(X \cap H_i)$).

Now on to the subtler case of the cubic. The whole theory here revolves around the astonishing fact that a smooth cubic always contains 27 lines—never more and never less and always intersecting each other in the same configuration! This can be shown as an application of (3.25) just like Bezout's theorem. However, before we can set this up we must construct auxiliary spaces to parametrize i) the smooth cubics in $\mathbb{P}^3$, and ii) the lines in $\mathbb{P}^3$.

(8.17) **The space of cubics.** The cubic forms

$$\sum_{|\alpha|=3} c_\alpha X_0^{\alpha_0} X_1^{\alpha_1} X_2^{\alpha_2} X_3^{\alpha_3}$$

have 20 coefficients c_α. Let $\mathbb{P}_W^{19}$ be the projective space with homogeneous coordinates W_α, all $\alpha = (\alpha_0, \alpha_1, \alpha_2, \alpha_3)$, $|\alpha| = 3$. For all $c = (c_\alpha) \in \mathbb{P}_W^{19}$, let H_c be the cubic surface $V(\sum c_\alpha X^\alpha)$ in $\mathbb{P}^3$. Then

$$\mathscr{H} \underset{def}{=} \bigcup_{c \in \mathbb{P}_W^{19}} H_c \subset \mathbb{P}^3 \times \mathbb{P}_W^{19}$$

is defined by the equation $\sum W_\alpha X^\alpha = 0$, hence is a closed algebraic set—in fact a variety of dimension 21. If $S \subset \mathscr{H}$ is the set of points where $p_2 : \mathscr{H} \longrightarrow \mathbb{P}_W^{19}$ is not smooth or equivalently where the fibre H_c is not smooth, then let $\Delta = p_2(S)$. Then $\mathbb{P}_W^{19} - \Delta$ parametrizes the smooth cubics.

(8.18) **The space of lines.** If $l \subset \mathbb{P}^3$ is a line, take 2 distinct points $a = (a_0, a_1, a_2, a_3)$ and $b = (b_0, b_1, b_2, b_3)$ on l. The 6 2×2 determinants

$$p_{ij}(l) = a_i b_j - a_j b_i, 0 \leq i < j \leq 3$$

are independent of the choice of a and b, except for multiplication by a common scalar. Also $a \neq b$ implies that at least one p_{ij} is not 0. Let $\mathbb{P}_V^5$ be the projective space with homogeneous coordinates $V_{ij}(0 \leq i < j \leq 3)$. The l determines a point

$$p(l) \in \mathbb{P}_V^5, \text{ by } V_{ij} = p_{ij}(l).$$

The p_{ij} are called the Plücker coordinates of l. It is easy to see that the p_{ij} determine l. In fact, if we define $\mathscr{Q}' \subset \mathbb{P}^3 \times \mathbb{P}_V^5$ by

(*)
$$\mathscr{Q}' \underset{def}{=} V\left(\begin{array}{c} V_{ij} X_k - V_{ik} X_j + V_{jk} X_i, \text{ for} \\ \text{all triples } 0 \leq i < j < k \geq 3 \end{array} \right)$$

then I claim:

$$\mathscr{Q}' \cap [\mathbb{P}^3 \times p(l)] = l.$$

Let W be the vector space of 4-tuples $(a_0, \ldots, a_3)$ of homogeneous coordinates; then more canonically, $(\ldots, p_{ij}, \ldots)$ is the point $a \wedge b \in \Lambda^2 W$ and the equations (*) are just $a \wedge b \wedge X = 0$ (in $\Lambda^3 W$). The solutions of this are $X = \lambda a + \mu b$, i.e., the points of the line $\overline{ab}$. In this notation, it is easy to verify that an element $p \in \Lambda^2 W$ is equal to $a \wedge b$ for some $a, b \in W$ if and only if $p \wedge p$ in $\Lambda^4 W$ is 0, i.e.,

$$p_{01} p_{23} - p_{02} p_{13} + p_{03} p_{12} = 0.$$

Let $G \underset{def}{=} V(V_{01} V_{23} - V_{02} V_{13} + V_{03} V_{12}) \subset \mathbb{P}_W^5$, and let $\mathscr{Q} = \mathscr{Q}' \cap (\mathbb{P}^3 \times G)$. Then the points of G are in one-one correspondence with the lines of $\mathbb{P}^3$ and $\mathscr{Q} \subset \mathbb{P}^3 \times G$

is the family of all lines. More notation: if $p \in G$, let l_p be the corresponding line. We will also need affine coordinates on G. Note that if $l \subset \mathbb{P}^3$ is a line:

$$[l \cap V(X_0, X_1) = \phi] \Leftrightarrow [p_{01}(l) \neq 0] \Leftrightarrow [p(l) \in G - G \cap V(V_{01})].$$

Look at the affine set of lines l such that $p_{01}(l) \neq 0$. In this set

$$\frac{V_{02}}{V_{01}}, \frac{V_{03}}{V_{01}}, \frac{V_{12}}{V_{01}}, \frac{V_{13}}{V_{01}}, \frac{V_{23}}{V_{01}}$$

are coordinates and

$$\frac{V_{23}}{V_{01}} = \frac{V_{02}}{V_{01}} \cdot \frac{V_{13}}{V_{01}} - \frac{V_{03}}{V_{01}} \cdot \frac{V_{12}}{V_{01}}$$

is the equation of G. Therefore this open set is isomorphic to $\mathbb{C}^4$. On the other hand, if $l \cap V(X_0, X_1) = \phi$, then $l \cap V(X_1)$ is one point with coordinates $(1, 0, a_2, a_3)$ and $l \cap V(X_0)$ is one point with coordinates $(0, 1, b_2, b_3)$. Then the affine coordinates of l are:

$$\frac{p_{02}}{p_{01}} = b_2 \; ; \frac{p_{03}}{p_{01}} = b_3 \; ; \frac{p_{12}}{p_{01}} = -a_2 \; ; \frac{p_{13}}{p_{01}} = -a_3.$$

In other words, the affine Plücker coordinates of l are just the affine coordinates of the points of intersection $l \cap V(X_0), l \cap V(X_1)$. Almost all of this generalizes to linear spaces L in $\mathbb{P}^n$, any r and n: one can construct a projective variety G, the *Grassmannian*, whose points are in one-one correspondence with the L's in $\mathbb{P}^n$, covered by affine varieties isomorphic to $\mathbb{C}^{(r+1)(n-r)}$. In general though G will not be a quadric hypersurface, but rather a subvariety of projective space of higher codimension.

Let's return to the cubic. Define

$$I \subset G \times \mathbb{P}_W^{19}$$

$$I = \{(p, c) \mid l_p \subset H_c\}.$$

(8.19) **Proposition.** *I is a smooth 19-dimensional variety and $p_2: I \longrightarrow \mathbb{P}_W^{19}$ is smooth over $\mathbb{P}_W^{19} - \Delta$.*

Proof. Let's first work out local equations for I. Look in the affine piece $G - G \cap V(V_{01})$ corresponding to the lines $l(a_2, a_3, b_2, b_3) = \overline{(1, 0, a_2, a_3)(0, 1, b_2, b_3)}$. Then

$$l(a_2, a_3, b_2, b_3) \subset H_c \Longleftarrow \sum c_\alpha X^\alpha \equiv 0 \text{ on all combinations}$$

$$s(1, 0, a_2, a_3) + t(0, 1, b_2, b_3)$$

$$\Longleftarrow \sum c_\alpha s^{\alpha_0} t^{\alpha_1} (sa_2 + tb_2)^{\alpha_2} \cdot (sa_3 + tb_3)^{\alpha_3} \equiv 0$$

Now expand this polynomial:

$$\sum c_\alpha s^{\alpha_0} t^{\alpha_1} (sa_2 + tb_2)^{\alpha_2} \cdot (sa_3 + tb_3)^{\alpha_3} \equiv s^3 \cdot f_0(a, b, c) + s^2 t f_1(a, b, c) + st^2 f_2(a, b, c)$$

$$+ t^3 f_3(a, b, c)$$

Then:

$$l(a_2,a_3,b_2,b_3) \subset H_c \Leftrightarrow f_i(a,b,c) = 0, 0 \le i \le 3.$$

Thus I is a closed algebraic set all of whose components have codimension ≤ 4 in $G \times \mathbb{P}^{19}_W$, i.e., dimension ≥ 19.

Next, look at the projection $p_1 : I \to G$. For any $l \subset \mathbb{P}^3$, the set of all cubic forms F that vanish identically on l is a *linear* subspace of the vector space of all cubic forms. Since all l's are projectively equivalent, the dimension of this linear space is independent of l. But if $l = V(X_0, X_1)$, then

$$F \equiv 0 \text{ on } l \Longleftrightarrow F = \sum_{\substack{\alpha = (\alpha_0, \alpha_1, \alpha_2, \alpha_3) \\ \alpha_0 + \alpha_1 \ge 1}} c_\alpha X^\alpha$$

which is a 16-dimensional space. Therefore, for all $p \in G, p_1^{-1}(p)$ is a 15-dimensional linear subspace of $\mathbb{P}^{19}_W$. In particular $p_1^{-1}(p)$ is irreducible; since G is irreducible, this shows that I is irreducible. Moreover the 4 local equations defining I, namely f_0, f_1, f_2, f_3, are linear in the c variable. Since for all a,b, they define a linear subspace of the space of c's of codimension 4, they are independent linear equations. Therefore their differentials in the c-direction are independent and therefore I is smooth of codimension exactly 4 in $G \times \mathbb{P}^{19}_W$.

Finally look at $p_2 : I \to \mathbb{P}^{19}_W$. We now want to prove that the differentials of the f_i's in the (a,b)-direction (i.e., on $T_{p,G}$) are independent *if H_c is smooth*. Fix a point $(p,c) \in I$ with H_c smooth and choose coordinates in $\mathbb{P}^3$ such that $l_p = V(X_2, X_3)$. Then in affine coordinates (a_2, a_3, b_2, b_3) on $G, p = (0,0,0,0)$. Let $F(X) = \sum c_\alpha X^\alpha$. Compute derivatives:

$$\frac{\partial}{\partial a_2}\left(\sum s^{3-i} t^i f_i(a,b,c)\right)\bigg|_{a=b=0} = \frac{\partial}{\partial a_2} F(s,t,sa_2 + tb_2, sa_3 + ta_3)\bigg|_{a=b=0} = s \cdot \frac{\partial F}{\partial X_2}(s,t,0,0)$$

$$\frac{\partial}{\partial b_2} \quad '' \quad = \frac{\partial}{\partial b_2} \quad '' \quad = t \cdot \frac{\partial F}{\partial X_2}(s,t,0,0)$$

$$\frac{\partial}{\partial a_3} \quad '' \quad = \frac{\partial}{\partial a_3} \quad '' \quad = s \cdot \frac{\partial F}{\partial X_3}(s,t,0,0)$$

$$\frac{\partial}{\partial b_3} \quad '' \quad = \frac{\partial}{\partial b_3} \quad '' \quad = t \cdot \frac{\partial F}{\partial X_3}(s,t,0,0)$$

Now the differentials of the 4 functions f_i with respect to the 4 variables a_2, a_3, b_2, b_3 are dependent if and only if $\exists \lambda_2, \lambda_3, \mu_2, \mu_3$ such that

$$\lambda_2 \frac{\partial f_i}{\partial a_2}\bigg|_{a=b=0} + \mu_2 \frac{\partial f_i}{\partial b_2}\bigg|_{a=b=0} + \lambda_3 \frac{\partial f_i}{\partial a_3}\bigg|_{a=b0} + \mu_3 \frac{\partial f_i}{\partial b_3}\bigg|_{a=b=0} = 0$$

for all i.

This is the same as saying that the 4 derivatives $\dfrac{\partial}{\partial a_2}, \dots, \dfrac{\partial}{\partial b_3}$ of $\sum s^{3-i} t^i f_i(a,b,c)$

at $a = b = 0$ are dependent polynomials. By our calculation this means:

$$(\lambda_2 s + \mu_2 t) \cdot \frac{\partial F}{\partial X_2}(s,t,0,0) + (\lambda_3 s + \mu_3 t) \cdot \frac{\partial F}{\partial X_3}(s,t,0,0) = 0.$$

Since these 2 partial derivatives are homogeneous *quadratic* polynomials, the existence of such linear polynomials $\lambda_2 s + \mu_2 t$, $\lambda_3 s + \mu_3 t$ is equivalent to $\frac{\partial F}{\partial X_2}(s,t,0,0)$, $\frac{\partial F}{\partial X_3}(s,t,0,0)$ having a common root $(\alpha,\beta,0,0)$. Since $l_p \subset H_c$, we can write F as $F = X_2 G + X_3 H$, hence $\frac{\partial F}{\partial X_0}$ and $\frac{\partial F}{\partial X_1}$ are also zero at $(\alpha,\beta,0,0)$. Therefore $(\alpha,\beta,0,0)$ would be a singular point of $H_c = V(F)$, hence it does not exist.   QED

(8.20) Corollary. *All smooth cubics contain the same finite number of lines.*

In fact, not only is the *number* of lines on H_c independent of c, but so is the *configuration* of these lines. Namely, if $l_1, l_2 \subset H_c$, then either $l_1 \cap l_2 = \emptyset$ or l_1 meets l_2 transversely in one point. It is convenient to express this situation by a graph Γ_c:

    1. vertices of Γ_c = lines in H_c
    2. (2 vertices are joined by an edge) $\Longleftrightarrow$ (the corresponding lines meet)

Then Γ_c is independent of c too. To see this, let

$$I' = \{(p_1,p_2,c) \mid l_{p_1} \subset H_c, l_{p_2} \subset H_c\} \subset G \times G \times (\mathbb{P}_W^{19} - \Delta)$$

Then one proves as in (8.19) that I' is smooth over $\mathbb{P}_W^{19} - \Delta$, so it is a disjoint union of smooth components I'_i of dimension 19. One of these, say I'_0, is the set of points (p,p,c), $(p,c) \in I$. For all other i's, l_{p_1} is never l_{p_2}. Look at:

$$\mathcal{H}'_i = \{(x,p_1,p_2,c) \mid (p_1,p_2,c) \in I'_i, x \in H_c\}$$

$$L_{ij} = \{(x,p_1,p_2,c) \mid (p_1,p_2,c) \in I'_i, x \in l_{p_j}\}$$

Then $\mathcal{H}'_i$ is a smooth 21-dimensional variety whose fibres over I'_i are H_c's, and $L_{i1}, L_{i2} \subset \mathcal{H}'_i$ are smooth 20-dimensional subvarieties whose fibres over I'_i are lines. Now $p_2 : L_{i1} \cap L_{i2} \to I'_i$ is injective since 2 lines meet in at most one point. By (3.28), all components of $\dim L_{i1} \cap L_{i2}$ have dimension 19. Therefore either $L_{i1} \cap L_{i2} = \emptyset$, i.e., l_{p_1}, l_{p_2} never meet if $(p_1,p_2,c) \in I'_i$, or $L_{i1} \cap L_{i2} \xrightarrow{\approx} I'_i$, i.e., they always meet.

Now one cubic on which we can quickly find all the lines is:

$$H_0 = V(X_0^3 + X_1^3 + X_2^3 + X_3^3).$$

Up to a permutation of coordinates, every line in $\mathbb{P}^3$ can be written $X_0 = aX_2 + bX_3$, $X_1 = cX_2 + dX_3$, Substituting this in the above equation, we find that this line is on H_0 if and only if

$$a^3 + c^3 + 1 = 0$$
$$b^3 + d^3 + 1 = 0$$
$$a^2 b + c^2 d = 0$$
$$ab^2 + cd^2 = 0.$$

If $abcd \neq 0$, one sees easily that these equations are contradictory. Say instead that $a = 0$: one sees then that $d = 0, b = -1, -\omega$ or $-\omega^2$ and $c = -1, -\omega$ or $-\omega^2$ (where $\omega = e^{2\pi i/3}$). Summarizing, we get on H_0 the 27 lines:

$$X_0 + \omega^i X_1 = X_2 + \omega^j X_3 = 0, \qquad 0 \leq i,j \leq 2$$
$$X_0 + \omega^i X_2 = X_1 + \omega^j X_3 = 0, \qquad 0 \leq i,j \leq 2$$
$$X_0 + \omega^i X_3 = X_1 + \omega^j X_2 = 0, \qquad 0 \leq i,j \leq 2.$$

Thus all cubics have 27 lines on them and their configuration can now be worked out (but it is not easy to visualize).

We can now prove that all smooth cubics are rational surfaces. On H_0, we find the lines:

$$l_1 : X_0 + \ X_1 = X_2 + \ X_3 = 0$$

$$l_2 : X_0 + \omega X_1 = X_2 + \omega X_3 = 0$$
$$m_1 : X_0 + \ X_1 = X_2 + \omega X_3 = 0$$
$$m_2 : X_0 + \omega X_1 = X_2 + \ X_3 = 0$$

$$m_3 : X_0 + \ X_2 = X_1 + \ X_3 = 0$$

$$m_4 : X_0 + \omega X_2 = X_1 + \omega X_3 = 0$$

$$m_5 : X_0 + \omega^2 X_2 = X_1 + \omega^2 X_3 = 0$$

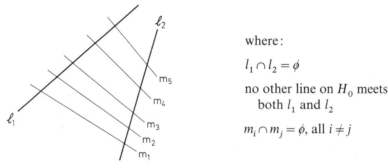

where:

$$l_1 \cap l_2 = \phi$$

no other line on H_0 meets both l_1 and l_2

$$m_i \cap m_j = \phi, \text{ all } i \neq j$$

By deformation, a similar set $(l_1, l_2, m_1, \ldots, m_5)$ of 7 lines exists on each cubic $H \subset \mathbb{P}^3$. In that case consider the projections:

$$\text{res } p_{l_i} : H - l_i \longrightarrow \mathbb{P}^1, i = 1, 2.$$

Now when projecting $\mathbb{P}^3$ from a line l, note that for all $\alpha \in \mathbb{P}^1, p_l^{-1}(\alpha) \cup l$ is a plane L^2 containing l and conversely if L is a plane containing $l, p_l(L - l) = \{\text{one pt. } \alpha\}$. Therefore the image space $\mathbb{P}^1$ can be identified canonically with the "pencil" of planes L^2 containing l. In particular, res $p_{l_i} : H - l_i \to \mathbb{P}^1$ maps a point $x \in H - l_i$ to the point of $\mathbb{P}^1$ corresponding to the plane $\overline{l_i x}$. But if $x_n \in H - l_i$ is a sequence of points such that $x_n \to x \in l_i$ as $n \to \infty$, then one sees easily that the sequence of planes $\overline{l_i x_n}$ tends to the plane $L = E^*_{x,H}$. Thus res p_{l_i} extends to an everywhere regular correspondence:

$$\overline{(\text{res } p_{l_i})} : H \longrightarrow \mathbb{P}^1.$$

Now look at the map:

$$\pi_1 : \overline{\text{res } p}_{l_1} \times \overline{\text{res } p}_{l_2} : H \longrightarrow \mathbb{P}^1 \times \mathbb{P}^1.$$

This is a regular correspondence with the following fibres:

(*)
$$
\begin{aligned}
&\text{If } \alpha \in \text{1st } \mathbb{P}^1 \text{ corresponds to } L_1 \supset l_1 \\
&\beta \in \text{2nd } \mathbb{P}^1 \text{ corresponds to } L_2 \supset l_2 \\
&\pi_1^{-1}(\alpha,\beta) = (H - l_1 - l_2) \cap L_1 \cap L_2 \\
&\text{plus the point } x = l_1 \cap L_2 \text{ if } L_1 \text{ tangent to } H \text{ at } x, \\
&\text{plus the point } x = l_2 \cap L_1 \text{ if } L_2 \text{ tangent to } H \text{ at } x.
\end{aligned}
$$

Let $m = L_1 \cap L_2$, $p_1 = l_1 \cap L_2$, $p_2 = l_2 \cap L_1$. m is a line meeting l_1 and l_2 at p_1 and p_2 respectively; conversely every line m meeting l_1 and l_2 is of the form $L_1 \cap L_2$ for a unique pair of planes L_i containing l_i. The following things can happen:

case i) $m \subset H$. Then m is one of the 5 lines m_i and $\pi_1^{-1}(\alpha,\beta) = m$.

case ii) $m \not\subset H$ and m meets H transversely at P_1 and P_2. Then by Bezout's theorem m meets H in exactly one point x besides p_1, p_2. By (*) $\{x\} = \pi_1^{-1}(\alpha,\beta)$.

case iii) $m \not\subset H$ but m is tangent to H at p_1. By Bezout's theorem, m meets H only in p_1 and p_2 and transversally at p_2, and by (*), $\pi_1^{-1}(\alpha,\beta) = \{p_1\}$.

case iv) $m \not\subset H$ but m is tangent to H at p_2. As in case iii), $\pi_1^{-1}(\alpha,\beta) = \{p_2\}$.

Therefore π is birational and has 5 exceptional curves m_i! Since $m_i \cap m_j = \phi$, the m_i's meet both l_1 and l_2 in 5 distinct points: therefore the points $Q_i = \pi_1(m_i)$ are not only distinct but no 2 of them lie on a single curve $\mathbb{P}^1 \times (\beta)$ or $(\alpha) \times \mathbb{P}^1$ of $\mathbb{P}^1 \times \mathbb{P}^1$. by (8.10) we have proven:

(8.21) Proposition. *For all smooth cubics H, $\exists$ a biregular correspondence*

$$\pi_1 : H \xrightarrow{\;\approx\;} B_{Q_1+Q_2+Q_3+Q_4+Q_5}(\mathbb{P}^1 \times \mathbb{P}^1)$$

for a suitable set Q_i of 5 points, no 2 on any "ruling" $\mathbb{P}^1 \times (\beta)$ or $(\alpha) \times \mathbb{P}^1$ of $\mathbb{P}^1 \times \mathbb{P}^1$

(8.22) Corollary. *For all smooth cubics H, $\exists$ a biregular correspondence*

$$\pi_2 : H \xrightarrow{\;\approx\;} B_{P_1 + \cdots + P_6}(\mathbb{P}^2)$$

for a suitable set P_i of 6 distinct points of $\mathbb{P}^2$.

Proof. Combine (8.21) with the biregular correspondence $B_{Q_5}(\mathbb{P}^1 \times \mathbb{P}^1) \cong B_{P_5+P_6}(\mathbb{P}^2)$ given by realizing $\mathbb{P}^1 \times \mathbb{P}^1$ as a quadric in $\mathbb{P}^3$ and projecting from Q_5.

The birational correspondence

$$\pi_2^{-1} : \mathbb{P}^2 \longrightarrow H \subset \mathbb{P}^3$$

has fundamental points $\{P_1,\ldots,P_6\}$, hence is defined by a 3-dimensional linear system Σ of curves with the 6 P_i as its base points. What is the degree of these curves? Let L be any plane in $\mathbb{P}^3$ meeting H in a smooth cubic curve (such an L is easily seen to exist). Let $C = L \cap H$. Now C meets l_i in points $x_i = l_i \cap L$. So the projection

$$\mathrm{res}\, p_{l_i} : C \longrightarrow \mathbb{P}^1$$

has degree $= [\deg C - \mathrm{mult}_{x_i}(C)] = 2$. Therefore $\pi_1(C)$ is a curve on $\mathbb{P}^1 \times \mathbb{P}^1$ whose projection to either factor has degree 2. Equivalently this means that we get transversal intersections $\pi_1(C) \cap (\alpha) \times \mathbb{P}^1$ and $\pi_1(C) \cap \mathbb{P}^1 \times (\beta)$ consisting of 2 points for almost all α and β. Now realize $\mathbb{P}^1 \times \mathbb{P}^1$ as a quadric K in $\mathbb{P}^3$. Then the curves $(\alpha) \times \mathbb{P}^1 \cup \mathbb{P}^1 \times (\beta)$ are particular plane sections of K and since these meet $\pi_1(C)$ transversely in 4 points, $\pi_1(C)$ has degree 4 in K. Now $\pi_2 = p_{Q_5} \circ \pi_1$, where p_{Q_5} is the projection of K to $\mathbb{P}^2$ from the point Q_5. Note that all the Q_i are in $\pi_1(C)$. Therefore $\deg(\pi_2(C)) = [\deg \pi_1(C) - \mathrm{mult}_{Q_5} \pi_1(C)] \leq 3$. If $x \in C - \cup (C \cap m_i)$, take a 2nd plane L' through x and let $C' = H \cap L'$. Then $\pi_2(C') \cap \pi_2(C)$ contains the 6 points $P_1, \ldots, P_6$ plus $\pi_2(x)$. Since

$$7 \leq \#(\pi_2(C) \cap \pi_2(C')) \leq \deg \pi_2 C \cdot \deg \pi_2 C'$$

it follows that $\deg \pi_2(C) = 3$. Therefore π_2^{-1} is defined by a linear system $\sum$ of cubic curves containing $P_1, \ldots, P_6$.

A consequence of this is that no 3 of the P_i are collinear: if P_1, P_2, P_3 were in a line l, then for all $D \in \sum$, either $D \cap l = \{P_1, P_2, P_3\}$ or $D \supset l$ since $\#(D \cap L) \leq 3$. But if $z, z' \in l - \{P_1, P_2, P_3\}$, then $\pi^{-1}(z) \neq \pi^{-1}(z')$, so $\exists L^2 \subset \mathbb{P}^3$ with $z \in L$, $z' \notin L$: then $\pi_2(H \cap L) = D \in \sum$ and $z \in \pi_2(H \cap L)$ but $z' \notin \pi_2(H \cap L)$ which is impossible. Next let V be the vector space of all cubic forms in X_0, X_1, X_2: counting coefficients you check that $\dim V = 10$. For any P_i, note that there is an $F_i \in V$ such that

$$F_i(P_i) \neq 0$$
$$F_i(P_j) = 0, \text{ all } j \neq i.$$

Just take F_i so that $V(F_i)$ equals 3 lines as follows:

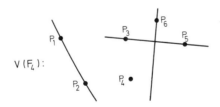

$V(F_4)$:

Therefore $\{F \in V \mid F(P_i) = 0, 1 \leq i \leq 6\}$ has co-dimension 6 in V, hence dimension 4, and the corresponding linear system has dimension 3. Since it contains $\sum$, it must equal $\sum$. We have now proven:

(8.23) **Proposition.** *Let H be a smooth cubic and*

$$\pi_2 : H \longrightarrow \mathbb{P}^2$$

the regular correspondence defined above. If $P_1, \ldots, P_6$ are the fundamental points of π_2^{-1}, then no 3 of the P_i are collinear and π_2^{-1} is defined by the linear system of all cubics through $P_1, \ldots, P_6$.

We will stop here although there is a great deal more which can be said about cubics. The methods used already should enable the reader to prove:

A) Check that not all of the P_i lie on a conic and that the 27 lines on H are
    i) the 6 exceptional curves of $\pi_2 : H \to \mathbb{P}^2$
    ii) the proper transforms on H of the 15 lines $\overline{P_i P_j}$
    iii) the proper transforms on H of the 6 conics through a subset of 5 of
      the P_i.

B) Let $P_1, \ldots, P_6 \in \mathbb{P}^2$ be any 6 points, no 3 collinear and not all on a conic. Let $\sum$ be the linear system of cubics through the P_i. Prove that $\sum$ defines a rational map from $\mathbb{P}^2$ to $\mathbb{P}^3$ whose image is a smooth cubic.

C) Investigate the configuration of the lines: for instance, let

$$G = \left(\begin{array}{l} \text{the group of permutations of the lines leaving fixed} \\ \text{the incidence relations.} \end{array} \right)$$

Then $\# G = 51{,}840$.

Somewhat deeper are:

D) G has a subgroup G_0 of index 2 which is simple and

$$G_0 \cong \left(\begin{array}{l} \text{gp. of } 4 \times 4 \text{ unitary matrices over the field } F_4 \\ \text{with 4 elements and conjugation } \bar{x} = x^2 \end{array} \right) \bigg/ \begin{array}{l} \text{modulo scalar} \\ \text{matrices } \alpha . I_4 \end{array}$$

[This can be proven by considering the cubic $\sum\limits_{i=0}^{3} x_i^3$ over F_4 as a Hermitian form $\sum x_i \bar{x}_i$, hence the cubic $V(\sum x_i^3)$ in a characteristic 2 $\mathbb{P}^3$ is left fixed by the above unitary group.]

E) $G \cong$ Galois group of least normal extension of $\mathbb{C}(I)/\mathbb{C}(\mathbb{P}_W^{19}) \cong$ group of permutations of the lines on H_{co} obtained by carrying the 27 lines around in a loop in $\mathbb{P}_W^{19} - \Delta$.

Good references are: Baker [1], vol. 3, Ch. 4, and vol. 4, Ch. 4 and Miller, Blichfeldt, Dickson, [1], Ch. 19.

Bibliography

A.M.S. = American Math. Society

  [1] *Algebraic Geometry, Arcata*, 1974. Proc. of Symposia in pure Math. **29**. Providence 1975.

A-M = Atiyah, M., MacDonald, I.

  [1] *Introduction to Commutative Algebra*, Addison-Wesley, 1969.

Baker, H.

  [1] *Principles of Geometry*. In 6 volumes. Cambridge Univ. Press 1929–1940.

Bott, R.

  [1] *On a theorem of Lefschetz*. Michigan Math. J. **6**, 211–216.

Bourbaki, N.

  [1] *Algèbre Commutative*. Hermann 1964–69; English translation: Addison-Wesley 1972.

Coolidge, J.

  [1] *A treatise on algebraic plane curves*. Oxford Univ. Press 1931.

E.G.A. = *Eléments de la géométrie algébrique* by A. Grothendieck and J. Dieudonné. Publ. of the
  I.H.E.S. 4, 8, 11, 17, 20, 24, 28, 32; 1960–1967.

Greenberg, M.

  [1] *Lectures on algebraic topology*. Benjamin 1967.

Gunning, R.

  [1] *Lectures on complex analytic varieties*, Princeton Univ. Press, 1970.

G-R = Gunning, R., Rossi, H.

  [1] *Analytic Functions of Several Complex Variables*. Prentice-Hall 1965.

Hensel, K., Landsberg, G.

  [1] *Theorie der algebraischen Funktionen*. Leipzig 1902. Chelsea reprint 1965.

Hervé, M.

  [1] *Several Complex Variables*, Tata Studies in Math, Oxford Univ. Press, 1963.

Hironaka, H.

  [1] *Resolution of singularities of an algebraic variety over a field of characteristic zero*. Annals of Math.
  79, 109–326 (1964).

Kleiman, S.

  [1] *Hilbert's 15th Problem: Rigorous Foundation of Schubert's Enumerative Calculus*. To appear in
  A.M.S. Proc. of Symposium on Hilbert's Problems, 1974.

Lang, S.

  [1] *Algebra*. Addison-Wesley 1965.

  [2] *Analysis* II. Addison-Wesley 1969.

  [3] *Elliptic functions*. Addison-Wesley 1973.

Massey, W.

  [1] *Algebraic Topology—an Introduction*. Harcourt, Brace, 1967.

Miller, G., Blichfeldt, H., Dickson, L.

  [1] *Theory of finite groups*. Wiley 1916.

Milnor, J.

  [1] *Morse theory*. Annals of Math. Studies **51**. Princeton Univ. Press 1963.

  [2] *Singular points of complex hypersurfaces*. Annals of Math. Studies **61**. Princeton Univ. Press 1968.

  [3] *Topology from the Differentiable Viewpoint*. Univ. of Virginia Press 1965.

Mumford, D.

  [1] *Abelian Varieties*. Tata studies in Math. Oxford Univ. Press. 2nd edition 1975.

  [2] *Curves and their Jacobians*. Univ. of Michigan Press 1975.

Munkres, J.

[1] *Elementary differential topology.* Annals of Math. Studies **54**. Princeton Univ. Press 1963.

Narasimhan, R.

[1] *Analysis on real and complex manifolds.* North-Holland 1968.

Šafarevič, I.

[1] *Algebraic surfaces.* Proc. Steklov Inst. **75** (AMS translation 1967).

[2] *Basic Algebraic Geometry.* Springer-Verlag 1975.

Samuel, P.

[1] *Méthodes d'algèbre abstraite en géométrie algébrique.* Ergebnisse der Math. **4**. Springer-Verlag 1955.

Semple, J., Roth, L.

[1] *Introduction to Algebraic Geometry.* Oxford Univ. Press 1949.

Serre, J.-P.

[1] *Groupes algébriques et corps de classe.* Hermann 1959.

Severi, F.

[1] *Vorlesungen über algebraische Geometrie.* Teubner, Leipzig 1921.

Singer, I., Thorpe

[1] *Lecture notes on elementary topology and geometry.* Scott-Foresman 1967.

Spivak, M.

[1] *Calculus on manifolds.* Benjamin 1965.

Stoizenberg, G.

[1] *Volumes, limits and extensions of analytic varieties.* Springer Lecture Notes **19**. Springer-Verlag 1966.

van der Waerden, B. L.

[1] *Algebraische Geometrie.* Springer-Verlag 1938.

[2] *Modern Algebra* (transl. from the German). Ungar 1970.

Walker, R.

[1] *Algebraic Curves.* Princeton Univ. Press 1950 (reprinted: Dover 1962).

Warner, F.

[1] *Introduction to manifolds.* Scott-Foresman 1971.

Weil, A.

[1] Variétés Kähleriennes. Hermann 1958.

Weyl, H.

[1] *Die Idee der Riemannschen Flächen.* Teubner 1955.

Zariski, O.

[1] *Algebraic surfaces.* Ergebnisse der Math. **61**. Springer-Verlag, 2nd ed., 1971.

[2] *Collected Papers.* MIT Press. Vol I. 1972.

Z-S = Zariski, O., Samuel, P.

[1] *Commutative Algebra.* 2 volumes. Van Nostrand 1959. Reprinted by Springer-Verlag 1975.

List of Notations

1. Set-theoretic notations:
   ϕ empty set
   p_i projection onto ith factor
   $\#$ cardinality of a set
   Γ_f graph, in $X \times Y$, of a map $f : X \to Y$

2. Algebro-geometric notations introduced in text:

Index

Die Grundlehren der mathematischen Wissenschaften
in Einzeldarstellungen
mit besonderer Berücksichtigung der Anwendungsgebiete

Eine Auswahl